Negation, Text Worlds, and Discourse:
The Pragmatics of Fiction

Advances in Discourse Processes

Roy O. Freedle, Series Editor

Negation, Text Worlds, and Discourse: The Pragmatics of Fiction

by Laura Hidalgo Downing

 Ablex Publishing Corporation
Stamford, Connecticut

Printed in the United States of America

Library of Congress Cataloguing-in-Publication Data

Hidalgo Downing, Laura.
   Negation, text worlds, and discourse: the pragmatics of fiction / by Laura
Hidalgo Downing.
      p. cm. — (Advances in discourse processes ; vol. 66)
   Includes bibliographical references and index.
   ISBN 1-56750-474-4 (cloth) — ISBN 1-56750-475-2 (pbk.)
   1. Discourse analysis, Literary. 2. Grammar, Comparative and general—
Negatives. 3. Heller, Joseph. Catch-22. 4. Negation (Logic). 5. Literature—
Philosophy. 6. Style, Literary. I. Title. II. Series: Advances in discourse
processes ; v. 66.
P302.5.H352000
808'.001'4—dc2199–36412
CIP

Ablex Publishing Corporation
100 Prospect Street
P.O. Box 811
Stamford, Connecticut 06904-0811

P

To my parents,
from whom I learned about language, logic, and humor,

and to the memory of Paul Werth

Contents

Acknowledgments

This book is a revised version of my Ph.D. thesis, which was submitted in June 1997 at the Universidad Complutense of Madrid in Spain.

I thank Joseph Heller for granting me permission to reproduce extracts from the novel *Catch-22*.

I am very grateful to several persons for their help and support during the writing of this book. I thank the supervisor of my thesis, Dr. JoAnne Neff, for her patience, advice, and support, and for her encouragement of my work. I am very grateful for the long talks at particularly stressful moments for me, and to have her as head of the department.

I also thank other friends, such as Clara Calvo, Pepe Simón, Manuel Leonetti, Manuela Romano, Chris Butler, Ana Martín Uriz and Jesús Romero, for their critical readings of different chapters. I also thank friends and colleagues at my university for their support and help while writing this book, especially Rachel Whittaker and Ana Martín Uriz.

I am also grateful to my friends from the PALA Association and from the Systemic Workshops for the discussion on papers related to the present work.

I am especially grateful to the late Paul Werth, for his friendship and for his encouragement of my work. A great part of this book is inspired by his work, including his last book which I was privileged to read in manuscript form long before his death in 1995 and which was finally published in July 1999. I hope this book will be a humble contribution to the perpetuation of his memory and to the continuation of his work.

I am very grateful to my parents and family, especially to my mother, Angela Downing, and to my father, Enrique Hidalgo, for encouraging and supporting me day after day, and for their help and advice on aspects of this volume. To them I owe the intellectual challenge and motivation to ever start writing a book.

I also thank my closest friends for their emotional support, and especially for entertaining me during this period. The informal discussions about *Catch-22* have been particularly rewarding, especially those at *La tertulia*, with Antonio and the rest, including Isabel, who has always been there even when she wasn't.

I am indebted most to Miguel Treviño, who has advised me about aspects of the book and who has helped me get through the hardest parts with his understanding and love; to him I owe the necessary peace and happiness to have been able to carry out and finish this book.

ABBREVIATIONS USED

AW: Actual world
TW: Text world
SW: Subworld
TAW: Text actual world
TRW: Text reference world
APW: Alternate possible world
K-world: Epistemic knowledge subworld
W-world: Wish subworld
O-world: Obligation subworld
MOP: Memory organization packet
TOP: Thematic organization packet

Introduction

Negation and affirmation appear to be inexorably linked. Consciousness, so far as
we know it, appears to be a rhythm of affirmation and negation, a power of
asserting and denying, of constituting and deleting.
Language, also, is a relation between affirmation and negation.
The word (or sign) is a presence based on absences, having meaning only because
it distinguishes, contrasts and excludes. It seems impossible, therefore, to speak of
a genealogical order of negation alone, to define negation from a past or ancestor
of its own outside of its perpetual polarity of negation and affirmation.
Kurrick (1979, p. 1)

Negation may be viewed as a pun, a play upon the norm. It is used when—more
rarely in communication—one establishes the event rather than inertia as the
ground. On such a background, the non-event becomes—temporarily,
locally—more salient, thus more informative.
Givón (1993, p. 190)

The development of knowledge, including scientific research, is essentially a
process of filling in absences; of giving a voice to something that has not been said
before or that needs to be reformulated. In this sense, the positive act of affirmation,
of saying something, arises from the awareness of an absence and requires the (at
least partial) denial of what has been said before. The present work originates from
the need to develop an integrated dynamic model of negation in discourse that is
adequate for the understanding of the role of negation in an extensive and complex
piece of discourse, the novel *Catch-22*. Most of the work on negation is still strongly
influenced by traditional philosophical problems, such as the relation between
negation and presupposition and the supposed ambiguity of negation, but little work
has been carried out in the area of discourse. Approaches to negation within the
functional-cognitive tradition tend to focus on specific aspects of negation, for
example, its functions as a speech act or its cognitive properties. Few attempts have
been made to propose an integrated discourse model. Furthermore, studies of
negation, with a few exceptions, tend to be limited to brief selections of texts or to
invented examples, often isolated sentences or brief exchanges. This book fills a
gap in studies of negation in discourse by providing an up-to-date critical review
of the state of the art in negation and by proposing a model of negation that brings

together the semantic, cognitive, and pragmatic features of negation, which are crucial for an understanding of its role in discourse.

My interest in the topic of negation, and, in particular, negation in the novel *Catch-22* developed when analyzing more general pragmatic features of the discourse of the novel. It became obvious that so much was being said by consistently negating. This awareness was systematized as an intuition that negation is a foregrounded feature in the discourse of the novel *Catch-22* that contributes to the development of a specific world view. I thus set myself to the task of accounting for the role of negation in the discourse of this novel. Accordingly, the theoretical frameworks used in the present book can be distinguished with regard to two different levels of analysis, each corresponding to two different theoretical issues: (1) a general level of analysis that constitutes the background within the field of stylistics in which the discussion of negation as a linguistic feature and the textual analysis are situated; and (2) a more specific level of analysis regarding the theoretical frameworks adopted in the analysis of the functions of negation.

Chapter 1 is a discussion of the basic concepts in the stylistics tradition. Chapters 2 to 5 are devoted to a discussion of different approaches to negation, from the philosophy of language and logic to grammar and discourse-pragmatic theories. These chapters are organized as an exploration into the explanatory possibilities of various frameworks with the aim of reaching a description of a discourse model that is adequate for the description of negation as a discourse unit. In this sense, each chapter leads to the next in search of an adequate discourse pragmatic model of negation. The adoption of a final framework, however, does not involve the rejection of all those previously discussed. Rather, it is suggested that the text world model, which is introduced in Chapter 3 and further developed in Chapter 5, is particularly adequate for the description of negation as a discourse phenomenon, precisely because it incorporates the properties of negation which I consider to be crucial for the interpretation of negation in discourse, and in *Catch-22* in particular. These properties have to do with three aspects that have not received sufficient attention, or no attention at all, in previous works on negation: (1) the need for an integrated model of negation in discourse that takes as a point of departure its cognitive properties; (2) the dynamic dimension of discourse processing and understanding, with the consequent reformulation of the role of negation in discourse; and (3) the view of negation as a complex discourse process with an influence that may go beyond the clause. A model of negation that accounts for these features will also provide the necessary tools for an adequate analysis of phenomena such as contradiction and contrariety, which have traditionally been problematic.

This book consists of six chapters. Chapters 1 to 4 are essentially theoretical, Chapter 5 is an application of a text world model to the analysis and interpretation of the role of negation in the novel *Catch-22*. Each chapter, however, has a final section devoted to a discussion of the advantages and limitations of the models dealt with in that chapter.

Chapter 1 is an overview of the basic notions in stylistic analysis and of the aims of research based on stylistic principles. The final sections of the chapter focus on the novel *Catch-22*, and are intended to provide a background regarding the main themes of the novel, its situation within contemporary fiction, and the contributions made by the present work in contrast to previous works on *Catch-22*. Chapter 2 is an overview and discussion of negation according to different theories, from the philosophy of language and logic to psychology, grammar, and discourse pragmatic theories. Chapter 3 is a discussion of negation within the framework of text world theory, in particular Werth's (1995c) text world model and Ryan's (1991b) analysis of conflict in fictional worlds. Chapter 4 is a discussion of negation within a schema theoretic perspective and deals with contributions made in the fields of frame semantics and schema theory to the understanding of negation and contradiction, of humor and of literature. Chapters 1 to 4 provide the theoretical background discussion of key issues in the fields of negation and stylistic analysis, Chapter 5 is of an applied nature, and constitutes my contribution to research on negation. Chapter 5 consists of a proposal of an integrated text world model of negation in discourse. It is also an application of the theoretical concepts from text world theory to the analysis of the discourse functions of negation in the novel *Catch-22*.

Since the size of the novel precludes an analysis of each extract in detail, I have chosen illustrative examples for each section, and have suggested generalizations regarding the applicability of observations made in each section to other groups of examples and to themes of the novel as a whole. This means that, ideally, the observations made in each of the sections should be understood to apply to significant aspects of the novel as a whole, and not to the analysis and interpretation of isolated passages. Similarly, although the discussion of the present book is organized around the analysis of the role of negation in a specific novel, I assume that the model of negation proposed in Chapter 5 should be applicable to the analysis of other discourse types. The conclusions and remarks on further research are presented in Chapter 6, which is followed by an Appendix that consists of three figures.

1

Basic Concepts

There was only one catch and that was Catch-22, which specified that a concern for one's safety in the face of dangers that were real and immediate was the process of a rational mind. Orr was crazy and could be grounded. All he had to do was ask; and as soon as he did, he would no longer be crazy and would have to fly more missions.

(Heller, 1961, p. 62)

This chapter is a brief introduction to the objectives of stylistic analysis and to key concepts in views of literariness in current linguistic theory. As such, the present chapter provides the background against which the analysis of negation as a discourse phenomenon should be understood. The review of approaches to literariness is not meant to be exhaustive, but, rather, it highlights some of the most significant steps in the development of linguistic stylistics in recent decades. This chapter also introduces concepts that are crucial for a dynamic understanding of discourse processing and, consequently, of negation as a discourse phenomenon; finally, a brief introduction to the novel *Catch-22* is given.

1.1. THE OBJECTIVES OF STYLISTIC ANALYSIS

By stylistics, I am referring to a wide range of works carried out in the last two decades as part of a linguistic tradition whose main interest has been the analysis and interpretation of literary (and nonliterary) discourse. Strictly speaking, its origins are to be found in the work of the Russian formalists (see Shklovsky, 1917/1965) and, more particularly, of the Prague School structuralists (see Garvin,

1964) and, especially, Jakobson (see Jakobson, 1964), which is discussed in Section 1.2.

The main principle upon which stylistics is articulated is the idea that "the primary interpretative procedures used in the reading of a literary text are *linguistic* procedures" (Carter, 1982, p. 4). This view was already present in Jakobson (1964), who precisely pointed out that poetics should be considered as a branch of linguistics, since poetry is actually made of words, of language. Most work in stylistic analysis, however, does not claim to provide an "objective" or "scientific" interpretation of a literary text (see, for example, Carter, 1982; Leech & Short, 1981; Short, 1989; Toolan, 1992; Simpson, 1993). Rather, the aim of stylistic analysis is that of systematizing the intuitions that readers have of a literary work, making our comments less vague and impressionistic by using rigorous linguistic procedures as an instrument of analysis. As such, stylistics, rather than "dissection" of a text should be understood as observation that leads to insight and to a deeper understanding of the text in question (see Leech & Short, 1981, p. 5). The objectives of stylistic analysis are summarized by Short (1995, p. 53) as follows.

> its main aim is to explicate how our understanding of a text is achieved, by examining in detail the linguistic organization of the text and how a reader needs to interact with that linguistic organization to make sense of it. . . . But the main purpose of stylistics is to show how interpretation is achieved, and hence provide support for a particular view of the work under discussion.

The claim inherent in stylistics, that it is possible to establish a connection between linguistic form and interpretation, is what makes it the target of criticism from other theories, ranging from generative linguistics to reader-response theory. In approaches to interpretation there are extreme positions such as structuralist interpretations, now unanimously rejected, where form is actually identified with content (Jakobson, 1964; Mukarowsky, 1964), and, at the other extreme, theories that dissociate all meaning from the text itself and place it in the individual reader alone (Fish, 1980). Although both positions are equally radical, it is possible to carry out an analysis that will take into consideration both linguistic features and the reader as part of the context in which literature is produced.

Recent works in functional linguistics incorporate a strong cognitive component in their models (see Givón, 1993; Semino, 1997; Verdonk & Weber, 1995; Werth, 1995c/1999), thus establishing a connection between cognitive processes that enable human beings to interpret reality and the production and understanding of language, and, more particularly, of texts. Ideally, what is needed is a framework of analysis that will make explicit the connections between linguistic and textual features, on the one hand, and knowledge, including semantic content and pragmatic inferencing, on the other, and how all of this is stored and processed by the reader. In this way, a particular interpretation of a text can be said to be justified on linguistic and cognitive grounds. Of course, the interpretation reached will be a possible one, among other possible interpretations. The fact that a connection may be established between the presence of linguistic features and their processing and interpretation by a reader does not guarantee that the resulting interpretation will be the same in different readers, or remain identical in the same reader during different readings of the work. Rather, a cognitive approach to text processing should be able to account for the natural tendency for interpretations to be *different* in different readers. Indeed, as linguists and critics have been arguing for some time now, the focus of discussions on texts should be on *how* interpretations arise, rather than *what* exactly is the interpretation of a given work. In fact, as readers, we know that our understanding and appreciation of a literary work can vary depending on different factors. This, however, does not invalidate the claim that specific evocations in the minds of readers may arise from specific linguistic aspects of the text.

The discussion of subsequent chapters explores to what extent current discourse theories can provide the analytical tools necessary for the study and interpretation of negation as a discourse phenomenon within a specific literary work, the novel *Catch-22*.[1]

1.2. THE IMPORTANCE OF TEXT: A COGNITIVE REFORMULATION OF THE NOTIONS OF *FOREGROUNDING* AND *DEFAMILIARIZATION*

As stated previously, the origins of stylistics as a discipline integrating linguistic analysis and an interest in poetics, can be traced back to the work of the Russian formalists (Shklovsky, 1917/1965) and, especially, of the Prague School linguists (Havránek, 1964; Jakobson, 1964; Mukarowsky, 1964). In this section, two concepts introduced by these linguists that are key aspects of studies in stylistics are discussed. Also, my view of literariness as part of the analysis of *Catch-22* is outlined. First, the notions of *defamiliarization* and *foregrounding* as linguistic phenomena that can help us understand the nature of literariness are developed. Second, the connection between these notions, particularly that of defamiliarization, is made with the function that can be assigned to literature as a discourse type. The latter is a question we come back to in different chapters of this book.

Shklovsky (1917/1965) first introduced the concept of *defamiliarization* or *deautomatization* to refer to the process whereby things are perceived anew, in a different light, which he illustrated with his famous example of discovering the "stoniness" of a stone. According to Shklovski, the purpose of poetic language should be precisely that of defamiliarizing taken-for-granted experiences. Similarly, other authors have since described the deautomatizing function of literature in different ways (see Burton, 1980; Garvin, 1964; Pratt, 1977). Burton (1980) discusses this concept with regard to its relevance for modern drama and reflects that "modern literature can be seen to be in a self-consciously designed Alienation tradition, whose central aim is to shock and disturb" the complacent audience

(Burton, 1980, p. 111). This issue and its more direct significance for 20th-century literature is discussed in Section 1.6.

Although the Prague School linguists in general claim that all varieties of language can present foregrounded features and deviance, most of their interest is directed toward exploring to what extent literature presents deviations from the standard. In this line, Havránek (1964) explores the notion of automatization and deautomatization and suggests that all varieties of languages have automatizing devices, which to a certain extent are shared with other varieties. Automatization is violated by means of *foregrounding* certain linguistic features, a device which deautomatizes the reading process and is described by Havránek (1964) as follows: "By foregrounding . . . we mean the use of the devices of the language in such way that this use attracts attention and is perceived as uncommon." Mukarowsky (1964) and Jakobson (1964) explore the functions of foregrounding in poetic texts, claiming that a distinguishing feature of poetic language is that it presents consistent and systematic foregrounded features that constitute a deviance from the norm in the standard. In Mukarowsky (1964, p. 19), we find a concise definition of the terms in question and their relation to poetics.

> The function of poetic language consists in the maximum of foregrounding of the utterance. Foregrounding is the opposite of automatization, that is, the deautomatization of an act; . . . Objectively speaking: automatization schematizes an event; foregrounding means the violation of the scheme.

Here, as in Jakobson (1964), a clear correspondence is established between poetic function and the presence of foregrounding and deviation from a norm.

The notions of foregrounding and deviation are taken up by Halliday, who argues that "foregrounding, as I understand it, is prominence that is motivated" (Halliday, 1973, p. 112), that is, related to the main theme, or one of the main themes of the work. This means that foregrounding itself does not guarantee the presence of a particular effect, and for this reason, it needs to be consistent and systematic. Furthermore, Halliday (1973, p. 115) modifies the original structuralist view that foregrounding and deviation are exclusively qualitative phenomena and suggests that foregrounding and deviance can also be quantitative. From this perspective, a feature might be foregrounded and produce deviant language, not because its presence constitutes a deviation from the use in the standard, but rather because, for reasons of frequency of distribution, the given feature is more or less frequent than would be expected. Halliday (1973, p. 115) calls this variant of deviation *deflection*.[2]

It is precisely the idea of the violation of a norm, which is the basis for the notion of deviation, which turns out to be problematic, as it becomes difficult to talk about one single norm from which a given piece of language can deviate, be it in qualitative or quantitative terms (see, for example, Carter & Nash, 1990, p. 5). Similarly, deviant language need not be present exclusively in literary texts (see

Carter & Nash, 1990; Cook, 1994; especially for deviance in advertisements), and, consequently, it becomes difficult to define literariness exclusively on the grounds of foregrounded features and deviant language. More serious is the criticism that the concept of deviance presupposes a reactionary ideology, since it seems to be based upon the assumption that there is a norm, defined in terms of positive attributes (standard, acceptable, or normal), from which other variants deviate. This view has as a consequence the definition of deviant texts as more marginal, less standard, less acceptable, and so forth. The advantage of working with a text model that does not make use of the notion of deviation is that variation and differentiation between texts are assumed to be natural phenomena, rather than deviations from a standard. In an approach of this type then, it becomes more important to describe the cognitive processes that give rise to different interpretations, as mentioned in Section 1.1. Indeed, in recent cognitive theories, a point is made of the central role played by linguistic acts such as storytelling and poetry, which are subject to the same type of cognitive processes as other linguistic acts (see Fauconnier, 1985, p. 1). These reasons, together with the fact that deviation as understood by the structuralists has in fact become the norm in literature (and art in general terms) of the second half of the 20th century, have led most critics to reject the notion of deviation, because of its inadequacies when trying to account for the process of literary interpretation.

In spite of their merit in directing the focus of attention of the interpretation process to the text itself, the work of the formalists has been criticized further for the claims implicit in Jakobsonian stylistics, this is, that the poetic function can in some way be identified with the presence of formal features in a text, such as parallelism. In Werth's work (1976), we find the first fully developed criticism of this point, and a suggestion by the author that formal analysis needs to be firmly grounded in semantics in order for the proposed analysis to be acceptable. Werth (1976, p. 65) argues that:

> to specify explicitly any effect associated with some linguistic feature configuration, we need three types of information . . . : (i) statistical, concerning the norms of language and its varieties; (ii) psychological, concerning the response of readers of different levels of sophistication to, perhaps, different paraphrases of the same content . . . ; (iii) semantic, concerning the thematic and implicational content of the given piece.

While point (i) in the quotation shows a concern for a "normative" view of language varieties, points (ii) and (iii) stress the need to take into consideration aspects that have been overlooked by the structuralists, namely, the response of readers and the semantics of the text.[3] In fact, the focus of attention of stylistics as a discipline has shifted naturally from the view of the text as a decontextualized product to a view that stresses the significance of cognitive and contextual factors in dynamic text processing.

The first author to explore the cognitive factors that determine the response of readers to books was van Peer (1986). In a study which was based upon a cognitive interpretation of the notion of foregrounding, van Peer provided empirical evidence for the connections between foregrounding and reader response, by means of a series of tests that checked reader reactions according to variables that could be used as parameters of foregrounding in literary texts (memorability, strikingness, importance, and discussion value). The results reveal that strikingness, importance, and discussion value support the theory of foregrounding, though memorability does not. In this way, an important redefinition of foregrounding is introduced, which is described by the author (van Peer, 1986, p. 20) in the following terms.

> Foregrounding, then, is to be understood as a pragmatic concept, referring to the dynamic interaction between author, (literary) text and reader. On the one hand, the material presence of certain foregrounding devices will guide the reader in his interpretation and evaluation of the text; on the other hand, the reader will look for such devices in order to satisfy his aesthetic needs in reading a literary text.

Van Peer's definition of foregrounding thus stresses the interactive dimension of the reading process. The significant contribution to the Jakobsonian theory of foregrounding is the cognitive reformulation of the concept, now understood in terms of the figure/ground distinction (van Peer, 1986, p. 21). Thus, foregrounded features are perceived by the reader as the figure, the foreground, or more salient information against the general background of the rest of the text. What becomes important in the process of interpretation is the relation between foreground and background, and how this is perceived by the reader in the reading process.

Van Peer's work has been followed by others on the same lines, among which it is worth mentioning Miall and Kuiken's (1994, 1998) studies on defamiliarization and the evocation of feelings and personal experiences in readers. This process is defined by the authors as follows.

> by defamiliarization we mean a process during which a reader uses prototypic concepts in a context where his or her referents are rendered unfamiliar by various stylistic devices; the reader is required to reinterpret such referents in nonprototypic ways, or even to relocate them in a new perspective that must be created during reading. (Miall & Kuiken, 1994, p. 337)

There is an obvious similarity between this definition and the one by Mukarowsky, with, however, a perspective that incorporates reader response in an explicit way; the structural notion of deviation is dropped in favor of the cognitive principle of prototypicality, so that the question is not whether a given text is deviant with respect to a norm, but whether concepts that arise in the reading of the text conform to prototypical models the reader brings into the reading process, or whether these models are in some way challenged. In this sense, the processing of literary texts takes place by means of the application of more general cognitive principles that govern understanding and interaction and are basically problem solving in nature (see, for example, Bernárdez, 1995, p. 37).

Following van Peer, Miall and Kuiken (1994, 1998) carry out experiments with readers that prove that there is a connection between the foregrounding of linguistic features and defamiliarization, on the one hand, and an effect of strikingness, higher difficulty in processing and evocation of personal experiences, on the other. Miall and Kuiken's results seem to confirm van Peer's claim that there actually is a cause–effect relation between linguistic foregrounding and reader response. Furthermore, the studies also show that the response will vary from reader to reader, depending on personal experiences, and, crucially, that feelings play a significant role (Miall & Kuiken, 1998, p. 338).

To sum up, the notions of foregrounding and defamiliarization are useful tools for the understanding of text processing from a cognitive stylistic perspective because they enable us to establish connections between linguistic features of the text and the cognitive processing of the text by readers. For this reason, I make use of these terms throughout this book.

1.3.   THE IMPORTANCE OF CONTEXT: LITERATURE AS DISCOURSE

As stated in Section 1.2., the merit of the formalists was to draw attention to the workings of the text itself in the process of literary understanding. As the discussion of the previous section illustrated, one of the most important criticisms to text theories has been precisely the failure to go beyond the structural analysis in order to consider a further dimension of meaning (see Cook, 1994; Halliday, 1973; Werth, 1976).[4] The first step toward developing a theory of literariness integrating levels of analysis not restricted to the text itself can be said to be found in two main trends of linguistic research; first, in work developed by systemic linguistic analysis as first outlined by Halliday (1973, 1978, 1994); second, in the contribution of the natural language philosophers (Austin, 1962; Searle, 1969) and subsequent pragmatic theories applied to literariness, which have given rise to interesting research.[5] In both these disciplines, the common feature is the consideration of literature as discourse, that is, as a text in context. This step obviously has important consequences, both from the point of view of the theoretic consideration of the nature of literariness and from the perspective of applications of methodology to the analysis and interpretation of literary books.

1.3.1.   The Hallidayan Approach to Literariness

By introducing the idea that literature is a discourse type, some significant consequences are derived. First, as is also claimed by cognitive theories, literature is basically not different from other discourse types, in that it can also be analyzed, from the perspective of the three main functions which can be found in any text, namely, the ideational, the interpersonal, and the textual functions (Halliday, 1973, 1978, 1994; Halliday & Hasan, 1985). As an illustration of this, Halliday provides a groundbreaking analysis of certain aspects of transitivity in W. Golding's book *The Inheritors*, which are described as deviations that yield an idiosyncratic view of the world, "a particular way of looking at experience" (Halliday, 1973, p. 120).[6]

Second, the consideration of literature from this perspective is demystifying with respect to the structuralist idea that literary texts are identified by the presence of a "poetic function." Systemic linguistics does not consider this to be one of the main functions of language; rather, it inverts the argument and claims that the main functions of language are also found in literature, as in other discourse types. This has the advantage of opening up the way for research from a wide variety of perspectives related to discourse issues.[7]

1.3.2. Pragmatic Approaches to Literariness

The view of literature as a discourse type makes literature available for analysis from pragmatic perspectives and for an analysis of the status of literature as a social phenomenon, aspects which are within the interests of systemic linguistics. Although the views of the natural language philosophers on literature were limited to a discussion of the nature of literature as a type of speech act,[8] their work has since been extended to the analysis of literary texts well beyond the limits initially envisaged by them. Pratt's (1977) extended development of a pragmatic theory of literariness, based on the Gricean notion of conversational implicature, was the first attempt in this direction, and has been followed by a large number of studies where pragmatic principles have been used as instruments of analysis of particular literary texts.[9]

To conclude, it can be said that the contributions of systemic linguistics and pragmatics to the notion of literariness as a type of discourse available for interpretation like other discourse types has been very influential as it has opened the way for research integrating different fields. For the purposes of this book, the consideration of literature as a discourse phenomenon is crucial, as it allows for interpretation at different processing levels, which include but also go beyond the text itself.

1.4. THE IMPORTANCE OF THE READER: LITERATURE AS INTERACTION

In the previous section, it has been made clear that a discourse approach to literariness requires the consideration of semantic and pragmatic aspects in addition to the formal features of the text itself. As Cook (1994, p. 44) points out, however, discourse and pragmatic theories focus on the significance of context from a social perspective and on inference procedures, and, consequently, show little interest in the role that the reader takes in the process of communication. An interest in the reader as active participant in the reading process has developed in linguistic theory, particularly with the emergence of dynamic approaches to text processing (see Carrell, Devine, & Eskey, 1988; Cook, 1994; Rumelhart, 1980; Schank, 1982; Schank & Abelson, 1977). Section 1.1. already pointed out how some authors (Miall & Kuiken, 1994, 1998; van Peer, 1986) have attempted to provide a cognitive basis for their studies on foregrounding and defamiliarization in the processing of literary texts, thus providing a place for reader response in the process of interpretation. Schema theories and text world theories, such as the ones described as follows, also address the issue of the cognitive aspects of discourse processing.

1.4.1. Schema Theory Approaches to Reading

The contribution of schema theory to theories of literariness will be discussed more in detail in Chapter 4. This introductory section provides a brief outline of those features of schema theory that are relevant in general terms when approaching literature as a dynamic discourse phenomenon where the reader takes an active part.

In the first place, schema theories claim that the procedures necessary for the understanding of a text can never be limited to the text itself, but rather, that cognitive features should be taken into consideration. As de Beaugrande (1987, p. 58) points out "There are no text properties in a vacuum: someone has to process the text and constitute the properties." Furthermore, it is this cognitive package that contributes to the creation of coherence as a discourse phenomenon, rather than specific links identifiable within the text (see van Dijk, 1977; Emmott, 1994; Rumelhart, 1980). Second, reading is seen as an active process of understanding that cannot be reduced to mere decoding. Reading involves active participation on the part of the reader, who maps the information in the text against her or his own stored knowledge to make sense of the text.[10] Finally, it is claimed that processing takes place at different levels of understanding, which range from understanding of discrete units involved in bottom-up processing to holistic understanding typical of top-down processing modes. (see, for example, Carrell, Devine, & Eskey, 1988; Rumelhart, 1980).

A discourse approach of this type is particularly interesting when applied to literature, precisely because of the possibility of dealing with the notions of dynamism and of different processing levels. The former, which has to do with the idea of text as a dynamic process rather than a static product, allows for an account of literariness where text understanding can be seen to change throughout the reading. This can help us identify interesting changes in works of fiction that are difficult to explain when considering the text as a static entity.[11] The latter, the idea of different processing levels in reading, allows the reader to interpret as meaning-ful, at a higher processing level, phenomena which are apparently contradictory, illogical, and meaningless at lower processing levels.

1.5. FICTIONS, WORLDS, TEXTS

The notion of processing at different levels of understanding becomes a necessary requirement when facing texts that are not ideal models of straightforward communication, but which make a point of being deliberately complex and often obscure. This is obviously the case of much literary discourse.

1.5.1. Literature and Ambiguity

The view of literature, as other art forms, as inherently ambiguous and paradoxical, is present in different theories and has been the object of discussion of philosophers, linguists, and literary theorists alike. The paradoxical nature of literature is related to the ambiguous relation that holds between fiction and reality.[12] To some extent, fiction reflects reality, at least *some kind* of reality recognizable by the reader. On the other hand, it creates a new reality, a new "world" (see Doležel, 1989; Iser, 1989; McHale, 1987; Ryan, 1991b). According to speech act theories (Searle, 1969, following Frege and Austin), this world, being fictitious, cannot be judged by the same rules that govern the "real" world, that is, it cannot be evaluated in terms of truth or falsity, since fiction lacks a referent in the real world. For this reason, fiction as a speech act is considered to be a "weak" or secondary type of act (see Petrey, 1990; Pratt, 1977; Searle, 1975).

In Bateson's (1972a) theory of play and fantasy, the weakened and paradoxical nature of literature has to do with the presence of what he calls a "metacommunicative message," *this is play.* This indicates that the acts carried out are not performed as denoting what they would otherwise do if it was not play. In this way, a paradox of the Russellian type is produced, in which what is denoted is not taken for what it should denote. The classical example of this kind of paradox is that of the Cretan liar saying "All Cretans are liars." Breuer (1980) applies Bateson's (1972) theory of play to literariness to argue that 20th-century literature is not only paradoxical in the terms described, but it reflects a split that is manifest in the widespread use of irony, which he compares to the split of the self in the schizophrenic person.

A fiction does have its own internal rules and conventions, however, which, as in any other discourse type, can be respected or violated (see Pratt, 1977; van Dijk, 1977). One of the idiosyncratic features of 20th-century literature is the tendency to deliberately break the conventions of fiction at different levels: within the text itself, thus creating what from a logical perspective are impossible worlds, or worlds which have internal contradictions, as in typical postmodern fiction; and at the level of interaction with the reader, by means of the defeat of expectations regarding notions such as what is fiction, what is poetry, and what is literature. Pratt (1977, p. 211) observes that:

In the literary speech situation, in other words, rule-breaking can be the point of the utterance. . . . Within literature, this kind of linguistic subversiveness is associated especially with the so-called "new" or "anti-novel," where we find radically decreasing conformity to the unmarked case for novels and a concomitant radical increase in the number and difficulty of implicatures required to make sense of a given text.

We find, then, that literature can be seen as paradoxical and conflictive from two complementary points of view. One, because as a type of art form it holds an ambiguous relation with reality; which it reflects but from which it is also different.[13] Second, because fiction is a discourse type with its own rules and conventions that are systematically challenged in 20th-century fiction. It has to be pointed out, however, that these violations are ultimately interpreted as meaningful, either through irony or other mechanisms that place a heavy burden of interpretation on the reader (see Breuer, 1980; Iser, 1989).

1.5.2. Possible Worlds and the Ontological Status of Fiction

Possible world theory as text theory (Dolezel, 1989; Pavel, 1986; Ryan, 1991b; Semino, 1995, 1997; Werth, 1995c) provides the means of dealing with the conflictive relationship between different "realities" by accepting that each world has its own internal configuration and laws. This view, summarized by Doležel, provides a solution to two traditional problems faced by philosophy with regard to the status of fiction: (1) what is the ontological status of fictional objects; and (2) what is the logical status of fictional representations (p. 221).

The problematic status of fiction as a nonexistent object is tackled by possible world theories by defining fictional worlds as possible states of affairs. Thus, Doležel (1989, p. 230) puts forward an account of fiction that develops from three main assumptions regarding the characteristics of fiction as a possible world.

1. Fictional worlds are possible states of affairs.
2. The set of fictional worlds is unlimited and maximally varied.
3. Fictional worlds are constructs of textual activity.

By means of these principles, we can deal with a fictional world as an alternative to the actual world with its own laws and internal structure. Furthermore, these characteristics are defined textually. This point is particularly interesting from the perspective of discourse-based approaches, as it provides the starting point for an analysis of fiction as constituting a semantic domain identifiable by means of linguistic properties. Ryan (1991b) takes these principles as a point of departure for the notion of *recentering*. The basic idea behind this notion is that every world is the center for its inhabitants. Thus, in reading *Hamlet*, one does so with the awareness that the *actual world* where one lives (which is only one, the "real world") is not perceived as such by the characters of the fictional world one is reading about. Thus, for the characters in *Hamlet*, the actual world will be the imaginary—though inspired in a historical time and place—world depicted in the play.

Possible world theories as text theories are an alternative to pragmatic theories based on speech acts and conversational implicature. They provide the means of considering a fictional world as a domain with its own rules and laws. They also provide the means of analyzing how conflict develops within the fictional world, from the perspective of the conflict between the domains of characters and the status quo in the fictional world (see Ryan, 1991b), or by observing how specific fictional types depart from accepted conventions or rules. To this respect, a possible world approach based on textual principles leads to the acceptance of what in logic would be defined as *impossible worlds*. Doležel (1989) makes the following observation with regard to the status of impossible worlds.

> Literature offers the means for constructing impossible worlds, but at the price of frustrating the whole enterprise: fictional existence in impossible worlds cannot be made authentic. The Leibnizian restriction is circumvented, but not cancelled. (p. 239)

Similarly, Eco (1989) rejects the status of impossible worlds (that is, worlds with internal contradictions) as fully-fledged worlds. Ryan (1991b), however, proposes a model based on a taxonomy of criteria that allow for the establishment of degrees of similarity with the real world and its laws and departures from it. This proposal has the advantage of working on the hypothesis that the differences between "actual," "possible," and "impossible" are a question of degree, rather than absolute distinctions.

The discussion of Chapter 3 further illustrates the advantages of approaching literary discourse from a text world perspective.

1.5.3.   Fictional Worlds as Text Worlds

In recent years, text world models have been proposed by different authors to account for the way in which text processing is also understood as "world creating" (see de Beaugrande, 1980; de Beaugrande & Dressler, 1981; Enkvist, 1989; Semino, 1994, 1997; Werth, 1995a, 1995b, 1995c). This view incorporates the active role of the reader, typical of dynamic approaches to text processing, to the notion of text as a possible state of affairs, and is based on the cognitive principle that there is no objective reality directly accessible to a human mind, but that each mind reconstructs a version of that reality according to each individual experience. Thus, Werth (1995b, p. 183), for example, argues that "There is a universe outside ourselves . . . but we can experience it only indirectly, by way of mental processing." With regard to text processing, the capacity to evoke imaginary situations through language is well known, and can be recognized as a typical feature of literary texts, as explained by Werth (1995b) as follows.

> By reading or listening to someone else's language, we can be transported mentally to situations experienced by other people, or even to entirely imaginary situations. Literature, of course, provides the best example of this (p. 184)

Similarly, Semino (1997, p. 57) observes that "Imagining worlds that are impossible by the standards of what we call the 'real' world is an activity we engage in (and enjoy) from childhood onwards." Thus, it can be argued that the reading process, like the process of interpreting and dealing with reality, involves the construction of a world or universe that is evoked by that text in our minds. To this aspect, Semino (p. 1) explains that "When we read, we actively infer a text world 'behind' the text." More specifically, Semino defines a text world as "the context, scenario or type of reality that is evoked in our minds during reading and that (we conclude) is referred to by the text." In the construction of the context evoked by a given text, the role played by certain elements such as deictic terms, frame knowledge, and metaphor are crucial (see Semino, 1997; Werth, 1995c); as is illustrated by the discussion of Chapters 3 and 4 which follow, the role of frame knowledge, in particular, is extremely important for the understanding of the function of negation within a text world model. For example, in Werth's (1995a, p. 78) discourse model, a text world is defined as follows.

> A world . . . is a conceptual domain representing a state of affairs. A text world, in particular, represents the principal state of affairs expressed in the discourse. First, the world must be defined; this is effected by means of the deictic and referential elements nominated in the text, and fleshed out from knowledge (specifically, knowledge frames).

The notion of text world outlined here is taken as a point of departure for the discussion of the role of negation in Chapters 3 to 5.

1.6.   LANGUAGE AND ACTION:
LITERATURE AND IDEOLOGY

The characteristics of the literary text, in which experiences from a shared reality are evoked in the mind of a reader, while at the same time, a "new" reality is created makes it a natural vehicle for ideological issues. *Catch-22* is a good example, in the sense that the highly idiosyncratic fictional world of *Catch-22* is at once recognizably fictitious and a mirror of very specific aspects of the society we inhabit, of which the novel is a satire.

This view of literature is grounded on the notion that language use is linked to a particular *world view* (Fowler, 1986) and that this world view reproduces an ideology. Fowler (p. 17) describes the process whereby language is used as a means of interpreting the real world and operating on it.

> It seems, then, that human beings do not engage directly with the objective world, but relate to it by means of systems of classification which simplify objective phenomena, and make them manageable, economical subjects for thought and action. In a sense, human beings create the world twice over, first transforming it through technology and then reinterpreting it by projecting classifications on to it.

The author then describes how the use of classification becomes naturalized in a given language, thus being taken as the "common-sense" view of reality, hypothesis, or ideology of the given community. In this sense, ideology is used to stand for a particular world view, which is reflected linguistically in the way in which language is used as a classifying tool. As is discussed in subsequent chapters, this notion is crucial when discussing the function of contradiction as a means of subverting or challenging established classifications.

Language classifies and gives shape to the reality we inhabit, thus conditioning our access to reality itself (see Hodge & Kress, 1994). Taken to an extreme, this view can lead, as in Bernstein, to a form of cultural determinism where the individual is trapped in the language system itself, which instead should be a tool enabling us to interpret and deal adequately with reality.[14] This view is particularly significant for an understanding of *Catch-22*, where language is used as a metaphorical trap into which the characters are systematically lured by higher authorities. In this sense, negation is a particularly effective instrument, in that it can easily lead to a form of "double-think," or the capacity to negate and affirm at the same time, if manipulated adequately (Hodge & Kress, 1994). This potentiality of negation as leading to double-think, that is, to negate and affirm at the same time, is accounted for in detail in the chapters on negation that follow.

Language reproduces an ideology, but it can also challenge it, and stylistic analysis can be used as a means of showing how manipulations of the linguistic system are related to ideologic manipulations (see Burton, 1982). As discussed previously, this implies a conception of literariness as one of the discourse types that can "do things with words," despite the fact that the natural language philosophers did not recognize this. Although this book is not directly concerned with ideology and critical discourse analysis, it is necessary to acknowledge the significance of this issue when dealing with a novel like *Catch-22*, which is obviously critical of a given status quo. Heller (in Krassner, 1993, p. 7) says about the subversive character of *Catch-22*:

> I think anything *critical* is subversive by nature in the sense that it does seek to change or reform something that exists by attacking it.

To sum up, the relationship between fictional world and reality in *Catch-22* is necessarily shaped as a distorted mirror image that operates as a criticism. Consequently, the discussion that follows in subsequent chapters deals indirectly with the significance of negation as a linguistic phenomenon that in some way contributes to the expression of criticism and satire.

1.7. *CATCH-22* AS AN ANTI-BELLUM NOVEL: NEGATION AND CONTRADICTION AS KEY CONCEPTS IN THE ANALYSIS OF THE NOVEL

The final section of this introductory chapter is intended to provide a general background to the discussion of the function of negation in *Catch-22* from the perspective of the general characteristics of the novel itself. The focus is on three main issues that are significant to the discussion and analysis of subsequent chapters: (1) a brief introduction to the cultural background against which the novel was created; (2) a summary of some of the main themes of the novel; and (3) an overview of the main studies on the novel.[15] This is also used as a way of introducing key themes that will be referred to throughout the book.

1.7.1. The Paranoid Hero in 20th-Century American Literature

The awareness of the cultural and historical context in which the novel was written becomes particularly necessary when talking about a work that is critical of a given situation. In an interesting series of appendices to his work *City of Words*, Tanner (1971) describes the social, historic, and psychological factors that characterize 20th-century American society and are typically present in its literature. In this section, we consider some of the aspects Tanner mentions that are relevant to the discussion of the present chapter.

Tanner explores the well-known assumption that 20th-century fiction, like other art forms, is the manifestation of a fragmented self alienated from the external world, which, at the same time, reflects the fragmented character of the external world itself.[16] As such, literature, according to its paradoxical nature described, both describes an external fragmented universe and an internal fragmented self, and within itself it is also fragmented as compared to art from previous centuries. In this sense, literature of the 20th-century is perhaps not strictly mimetic regarding its content, but rather, it has become iconic in the sense that its form imitates a given state of affairs (see Breuer, 1980). Heller makes the following observation about *Catch-22* (Krassner, 1993, p. 9):

> I tried to give it a structure that would reflect and complement the content of the book itself, and the content of the book really derives from our present atmosphere, which is one of chaos, of disorganization, of absurdity, of cruelty, of brutality, of insensitivity,

but at the same time one in which people, even the worst people, I think are basically good, are motivated by humane impulses.

Tanner makes use of the notion of the contemporary individual's incapacity to deal adequately with what he calls "available patterns of experience" (Krassner, 1993, p. 421). The dilemma is often present in the American hero, who has to choose between social values he does not conform to or his individuality. This clash between self and world is intensified by a feeling that the individual is predictable, manipulable, controllable. This view became widespread in the earlier 20th century as a result of scientific, psychological and social theories based on deterministic and behavioristic principles. Tanner (1971, p. 424) observes:

All this is anathema to the American hero, who will go to some lengths *not* to be what the situation seems to call for, in an attempt to assert his immunity from conditioning. Obviously this can produce a sort of negative determinism in which the non-conformist individual is predictably unpredictable.

This quotation is significant because it sheds light on the characteristics of Yossarian as the hero of *Catch-22*, precisely in the terms described, of his consistent attempt to contradict and sabotage the patterns of behavior imposed on him by external forces. As Seed (1989, p. 31) argues, Yossarian preserves his individuality through negative action. As a hero, Yossarian also shares the paranoid tendencies that are characteristic of 20th-century American fiction. This aspect is closely related to the perceived social situation described by Tanner (1971, p. 427) as involving "a vast conspiracy, plotting to shape individual consciousness to suit its own ends." This paranoid feeling is also present in other American heroes,[17] and it yields recurrent symbolic images representing a dread toward a Whorfian sociocultural determinism. Tanner (p. 429) provides an example from *Catch-22*, the image of the "white soldier," a soldier who is completely bandaged from head to foot, and immobilized. Throughout the novel, the characters begin to suspect that there is actually nobody inside the bandages, a realization that leads to a situation of panic in the hospital. Tanner defines the symbolism of this kind of image as the terror "of the individual 'as dynamically empty': the void under the bandages in *Catch-22*" (p. 429).

The American hero, however, does not actively provide new structures or solutions to the conflicts described. Tanner (1971) argues that there is seldom a reconstruction of the self in recent American fiction and there are no clear ideas that would substitute those against which the hero is rebelling. Again, Yossarian, as protagonist of *Catch-22*, fits this description, particularly in his final decision to ultimately escape from the untenable situation, rather than confront it directly.

To sum up, I have argued for a view of literariness as paradoxical and conflictive in a way that reflects the situation of the individual and his or her relation to society and art in the 20th century. This is a necessary background for the understanding of the general characteristics of recent American fiction and of the function of negation in the discourse of *Catch-22*.

1.7.2. Catch-22 and War Narrative

Catch-22 narrates the story of an American bombardier squadron on the imaginary island of Pianosa, off the Italian coast, during World War II. The novel is not meant to be realistic, in the sense that it does not faithfully reproduce the reality of World War II. Rather, World War II, as a prototypical example of a war, becomes the excuse to explore the workings of 20th-century society. Heller himself (Merrill, 1993, p. 160) points out that the novel has more to do with the situation of America during the Cold War, the Korean War, and the possibility of a Vietnam War, than with World War II. Hence, the implied criticism of the McCarthy era with the loyalty oaths, the trials, and the paranoid feeling toward the "non-American."

This is reflected in the novel mainly through a military system governed by highly inefficient bureaucrats, whose only concerns are the petty struggles for power against other military officers in the squadron or nearby squadrons. The war itself, or the enemy, are hardly present in the novel, again leading to a recurrent theme of recent American literature, that the enemy is found within the system itself, not outside. In this closed system, however, the victims are the ordinary soldiers, who are in the hands of the higher officers and their whims. Heller (Krassner, 1993, p. 22) emphasizes of *Catch-22*:

> I regard this essentially as a peacetime book. What distresses me very much is that . . . when this wartime emergency ideology is transplanted to peacetime, then you have this kind of lag which leads not only to absurd situations, but to very tragic situations.

Thus, the scope of the narration goes beyond the idea of criticizing war and, as said above, is used as an excuse to reflect on other current aspects of American society, such as religion, justice, morality, racism, and, ultimately, the power of institutions themselves. As Walsh (1982) indicates in a study of American war literature, *Catch-22* shares with contemporary fiction its concern with a dehumanized world "of indeterminate and anxious character" (p. 190).

> [Heller's] formal procedures such as the devices of satiric distortion, allegory, parody and burlesque contribute to the formation of a vision of breakdown. His novel exploits the departure of the monolithic power of modern institutions, in this case conveyed through the metaphor of the army's hierarchy.

According to Walsh, the novel also reflects the shift in focus in recent war fiction, which no longer deals with the battlefield but with "the absurdities of the communication process itself" (1982, p. 191). The catch-22 itself becomes the main metaphor for absurdity as the "seemingly willing mass subjection of soldiers to the interests of the industrial military complex" (Walsh, 1982, p. 190). Or, as another critic described it, "Catch-22 is a metaphysical principle of inbuilt chaos" (Hunt, 1974, p. 129). The aspects that *Catch-22* shares with typical war narratives, such as the idea of the military group as a home for one's loyalties, the mechanistic concept of existence, the relation between sexuality and war, and the religious overtones of war's sacrifice are all ridiculed in *Catch-22* (see Solomon, 1969).

1.7.3. The "Catch-22"

One of the most striking features of the novel *Catch-22*, in addition to its structure, is undoubtedly the peculiar logic that characterizes language use and exchanges.[18] Nash (1985, p. 110) attributes great part of the madness of the world of *Catch-22* to the nature of the fictional world and its communication exchanges as closed systems; thus, Nash (1985) argues that all characters in the novel are trapped "within a closed system of argument which envelops all, and from which they cannot escape because they have recourse only to propositions generated within the system." The catch-22 itself clearly summarizes this. Catch-22 states that a soldier who is crazy can be grounded, and can be sent home. In order to be grounded, he has to make an application; however, by applying to be grounded, the applicant will be proving he is able to use rational capacity and, consequently, that he is not crazy. The catch is a type of circular argument where two propositions (a) you need to be crazy (P), and (b): you need to apply (Q), cancel each other out (if you apply you are not crazy: *if P, then not-Q*), thus making it impossible for the proposition *to be grounded* ever to be applicable. The catch is a circular trap that can be explained as follows.

1. If you are crazy you can be grounded. *If A then B.*
2. If you want to be grounded you have to apply. *If C then D.*
3. If you apply you are not crazy. *If E then F.*

The process is circular because the last proposition (F) contradicts the first premise (A), so that $F = not\text{-}A$. At a literal level of understanding, the catch does not make sense. As a metaphor of the novel itself, however, it makes a lot of sense. It reflects the closed system that constitutes the fictional world from which there is practically no escape; it reflects the circularity present in the novel as a narrative device and as a systematic structuring principle for many exchanges between characters; it also reflects the blurring of opposites that are recurrent themes, such as craziness/sanity, presence/absence, life/death, thus challenging both language as a system that organizes experience, and experience itself. This leads to the closely linked modernist theme of the indeterminacy between reality and illusion, also present in the recurrence of images related to disturbances in visual perception, such as hallucinations, visions, dreams, nightmares, dejá vu and jamais vu (see Blues, 1971; Mellard, 1968).

1.7.4. Previous Works on *Catch-22*

Much of the literature on *Catch-22* has dealt with the themes mentioned previously (see Burnham, 1974; Blues, 1971; Davis, 1984; Gaukroger, 1970; Mellard, 1968; Nagel, 1974a, 1974b, 1984; Nelson, 1971; Pinsker, 1991; Protherough, 1971; Seed, 1989; Solomon, 1969; Walsh, 1982). Otherwise, it has tended to concentrate on the characteristics of the novel from the point of view of its narrative structure and chronology. The chaotic pattern of the novel has been attacked as lacking a structure (Waldmeir, 1964). It has also been defended as a chaotic structure intended to reflect the content of the novel (see Gaukroger, 1970). In this respect, different critics interpret this pattern in terms ranging from musical counterpoint, to information theory and entropy (Tucker, 1984), the theory of chaos (see Derks, Palmer, Safer, Sherman, & Svebak, 1994), the psychological theory of synergies (see Derks et al., 1994) or, more generally, as a cyclical pattern where relevant episodes are repeated incrementally (Protherough, 1971; Seed, 1989). Many critics agree in defining the narrative unit of the novel as the "episode" and also point out the importance of two time lines, which are the main reference points throughout the narrative: the psychological time of development and the linear time line of the increase in the number of missions (see Burnham, 1974; Gaukroger, 1970; Nagel, 1974a; Seed, 1989; Solomon, 1974).

Although the paradoxical nature of the novel is mentioned by many authors (Gaukroger, 1970; Greenberg, 1966; Ramsey, 1968; Seed, 1989), and negation is mentioned by Ruderman (1991), to my knowledge there is no single work devoted to the discussion of negation and contradiction in an extended way, be it an article or a book. Rather, most critics have concentrated on the absurdity conveyed through the use of circular logic (see Ruderman, 1991; Seed, 1989). The relationship between the absurd character of the novel and existentialism and absurd drama has also been pointed out in various places (Gaukroger, 1970; Ramsey, 1968). The present work is a contribution to the literature on *Catch-22* by focusing on negation and contradiction as linguistic phenomena.

1.8. CONCLUSION

In this introductory chapter, I argue for a view of literariness grounded on a cognitive interpretation of the notions of foregrounding and defamiliarization as useful tools for the analysis of negation and contradiction as natural foregrounding devices, which may lead to defamiliarization in a reading of *Catch-22*. This view is extended by a discussion of discourse-based approaches to literariness, arguing that the notion of context is essential when dealing with literary communication. This allows for the consideration of the crucial role of the reader in the reading process and in the construction of the fictional world evoked by the text. Literature has also been described as holding an ambiguous relation with reality, which in some way it reflects but from which it also differs.

The ambiguous nature of literature is used as the touchstone for the discussion of the relationship between fiction and the reality of the 20th century, and the relation of fiction with ideology. Within this frame, *Catch-22* is described as a novel that reflects its times, in that a paradoxical fictional world both reflects and criticizes a paradoxical and absurd real world. Greenberg (1966) describes "the novel of disintegration in the twentieth century"—of which *Catch-22* is an example—in the following way:

> The novel of disintegration, in its focus upon the paradox of individual involvement in the entropic process, thus emerges as specifically the literature of the human condition: a literature at once subversive (to the general tendencies of the world) and loyal (to the value of the individual), tragic (in its ultimate prognosis) and comic (in its present potential), which opens through human necessity to human possibility. (p. 124)

NOTES

1. The approach to negation followed in the present work is grounded on linguistic theory. For an approach to the notion of "negativity" in literary theory, see Budick and Iser (1989).

2. In Halliday (1973), in addition to the theoretical considerations regarding foregrounding and deviation, the author also presents his view on the functions of language and, finally, his groundbreaking work on transitivity in W. Golding's book *The Inheritors*.

3. Compare Leech and Short's (1981) discussion of the notions of foregrounding and deviance: for these authors, deviance is a statistical notion, prominence is a psychological notion, and foregrounding constitutes what is literary relevant. There is a close connection between the three different phenomena.

4. This is one of the limitations of stylistic analysis in the transformational-generative line, which in the earlier stages of stylistics was very popular. As pointed out in Section 1.2., however, studies such as van Peer (1986) and Miall and Kuiken (1994) do incorporate the reader in their models based on the notions of the Russian formalists.

5. See van Dijk (1977 and 1985) and de Beaugrande (1980) for different examples of text theories incorporating semantic and pragmatic principles to text analysis.

6. Compare Fowler's (1986, p. 44) notion of "world view." It is reproduced in Section 1.6.

7. See for example Widdowson (1975, 1992), Carter (1982), Carter and Simpson (1989), and Verdonk and Weber (1995) for discourse approaches to literary analysis; Fowler (1977, 1986), Burton (1982), Birch (1989), and Simpson (1993) for critical discourse approaches to literariness; Burton (1980) for a discourse theory of dramatic discourse.

8. See Petrey (1990) for a detailed discussion of the views of the natural language philosophers on literariness and suggestions for a modification of their perspectives.

9. See Simpson (1989) and Calvo (1992) for an interpretation of dramatic discourse in terms of Brown and Levinson's (1987) model of politeness, and Pilkington (1996) for an application of relevance theory to the interpretation of poetic discourse.

10. See also Iser (1989) for an approach to reading as dynamic and requiring reader involvement from a perspective based on game theory and inspired in Bateson (1972).

11. This view also has important repercussions in the methodology of teaching language and literature as integrated disciplines and based on a process-oriented approach, rather than a product-oriented one.

12. For a discussion of ambiguity as an inherent characteristic of literary discourse see Empson (1930/1995).

13. In this respect, Werth (1995c, p. 499) argues that fiction "is deemed to have a symbolic relationship with the world of experience."

14. Compare with Tanner's account of the social situation in 20th-century America.

15. See Weixlmann (1974) and Keegan (1978) for complete reference guides to Joseph Heller and *Catch-22*.

16. See also Davis (1984) for a development of the idea that 20th-century fiction after modernism reflects the discontinuities in language, thought, and society, which characterize this century.

17. See Bockting (1994) for an application of the notion of mind style to characterization in Faulkner. In this analysis, the author shows that Jason Compson's language reveals paranoid tendencies in his character.

18. Compare Nash (1985, pp. 110–111) for a comparison of the language in the exchanges in *Catch-22* with the language of psychiatric patients.

2

Approaches to Negation

Ordinarily, Yossarian's pilot was McWatt, who, shaving in loud red, clean pajamas outside his tent each morning was one of the odd, ironic, incomprehensible things surrounding Yossarian. McWatt was the craziest combat man of them all probably, because he was perfectly sane and still did not mind the war. . . . McWatt wore fleecy bedroom slippers with his red pajamas and slept between freshly pressed colored bedsheets like the one Milo had retrieved half of for him from the grinning thief with the sweet tooth in exchange for none of the pitted dates Milo had borrowed from Yossarian.
(Heller, 1961, p. 80)

2.1. INTRODUCTION

Negation is a challenging and difficult subject that has posed problems for philosophers, logicians, psychologists, and linguists alike. Within the study of language, the affirmative seems to be quite straightforward; negation, by comparison, is extremely difficult to define and describe. For this reason, researchers have long been troubled by apparently trivial questions, such as: what exactly is negation? What is the ontological status of negative entities? Is negation ambiguous? Does negation presuppose the affirmative? Can we talk about negative lexical items? and so on.

This chapter addresses some of these questions by reviewing works on negation in different fields, from the philosophy of language, logic and psychology, to grammar and discourse, although the main focus is on functional approaches in current linguistic theory. This initial analysis has the objective of showing how approaches based on traditional logic and semantics are not sufficient for an understanding of the function of negation in discourse, and in particular, of the discourse of *Catch-22*. This leads to the discussion of later chapters, where discourse-based frameworks are explored with regard to their adequacy for the understanding of negation.

2.2.   INTRODUCTORY NOTIONS ON NEGATION

In the subsequent sections, the crucial role of negation in the philosophical and psychological traditions as a background to the more linguistically oriented discussions is reviewed. These sections provide an illustration of the central position occupied by negation in the tradition of Western thought, and they also introduce concepts that have been extremely influential in the way negation has been understood in language.

2.2.1.   Negation in the Philosophy of Language Tradition

Negation is one of the major controversial issues discussed by philosophers, psychologists, and linguists for centuries. Even if it has always occupied a central position in studies of logic, as Horn (1989, p. 1) observes, it has always "been regarded as a suspect guest, if not a spy from the extralogical domains." Most of the books on the subject, however, are concerned with issues that derive from philosophic and logic problems of sentences where negation is involved. Only in recent years has there been an attempt to study the properties of negation in language use.

2.2.1.1.   Negation and presupposition

Among the traditional problems associated with the presence of negation, the assignment of a truth value to sentences containing nonreferential noun phrases (NPs) has been particularly thorny, and has given rise to a discussion in which the most famous linguists and philosophers of this century have had something to say. The problem, introduced by Russell (1905/1988), involved the well-known propositions in (2).

> (2)   a.   The King of France is bald.
>   b.   The King of France is not bald—because there is no King of France.

The dilemma has to do with the answer to two problematic questions.

1.   Since there is at present no King of France, because France is a republic, is sentence (2)a false, or is it neither true nor false, thus implying there is a truth-value gap?

2. How does one deal with (2)b? In (2)b the negative proposition negates the existential presupposition present in the affirmative proposition, thus contradicting the general principle that the presuppositions of a sentence are kept unchanged under negation.

This kind of negation, which denies the presuppositions of a proposition rather than the proposition itself, has been called *external, widescope, or marked* negation, to contrast it to *internal, narrowscope, unmarked* negation. Different linguists and philosophers have offered varied solutions to the Russellian problems. Austin (1962, p. 51) considered utterances such as (2)a "void" with regard to truth value, and, consequently, infelicitous. Similarly, other authors have pointed out that the problem established in abstract terms is actually much less of a problem when the sentence is seen in context, from a pragmatic perspective. Kempson (1975) was among the first to defend a pragmatic approach to presupposition and a solution of the problems involved in assigning truth values via Grice's (1975) conversational maxims and the notion of implicature.[1] Givón (1993, p. 196–197) argues that an utterance like *The king of France is not bald*, as in (2)b, is not generally used to express widescope negation in natural language, as this meaning is generally expressed by means of alternative structures that are not ambiguous. Typically, these involve some form of subject NP-negation, as in (3) (Givón, 1993, p. 197).

(3) a. There is no King of France.
     b. No king of France is bald.

Indeed, the most common type of negation seems to be verb phrase negation (see Givón, 1993, p. 197; Tottie, 1991, p. 46), which, as Givón points out, "typically excludes the subject, and thus applies only to the *verb phrase* (predicate)" (1993). The reason for this is that "in human communication definite subjects are *not* used to assert their existence. Rather, their existence (and shared knowledge of their identity) is *presupposed*, it is part of the background information and thus does not fall under the scope of assertion to begin with." (Givón, 1984, p. 325).

2.2.1.2. Negation and ontology

The problem of nonreferential NPs and negation, of course, has ontological implications. In fact, a long-standing argument has traditionally existed in philosophy regarding the ontological status of objects in general, and, more specifically, of nonobjects. The traditional questions have been (1) what objects are there? and (2) what objects exist? Some philosophers, such as Russell (see Marsh, 1988) and Quine (see Dummett, 1981) have defended a view where only objects that can be assigned reference, in addition to sense, can have ontological status and, consequently, can be considered to exist. On the opposite end, other philosophers, such as Meinong (see Russell, 1905/1988, p. 45) defend a view where any grammatically correct denoting phrase can stand for an object. This means that expressions such as *The present King of France* and *the round square* are objects. Russell observes that "it is admitted that such objects do not *subsist*, but, nevertheless, they are supposed to be objects" (Russell, 1905/1988, p. 45). For Russell, such objects infringe the Law of Non-Contradiction because they assert that the King of France exists and does not exist at the same time. As a solution to this problem, Russell proposes a distinction between *primary* and *secondary* occurrences of denoting phrases. According to this distinction, the proposition *The present king of France is not bald* can be interpreted as meaning *There is an entity X who is not bald* and is an instance of a primary occurrence, or it may mean *It is not the case that there is an entity who is bald*, which is an instance of a secondary occurrence. In the former interpretation, the proposition is false; and in the latter, it is true. This distinction has had an enormous influence on theories of negation and is related with the distinction between narrowscope and widescope negation mentioned previously. A further solution to the problem of nonreferential NPs is proposed by the theory possible worlds, which is discussed in Chapter 3 (see also the introductory section in Chapter 1, Section 1.5.2).

2.2.1.3. *Metalinguistic negation*
A related logical problem that arises with regard to negation is that of scalar implicatures involving negatives (see Atlas & Levinson, 1981, p. 32; Horn, 1989, pp. 382–392), as in (4).

> (4)   He doesn't have three children—he has four.

This kind of negation goes against the principle that a scalar value will imply all the lower values below itself on a scale. Thus in example (4), having four children implies having three. Strictly speaking, it is not logical to deny that one has three children if one has four. As in the previous case, Horn (1989, p. 384) deals with these cases as special instances of *metalinguistic negation*, where what is denied is not the content of the proposition but the proposition as a whole. Horn defines metalinguistic negation as follows.

> METALINGUISTIC NEGATION: a device for objecting to a previous utterance on any grounds whatever, including the conventional or conversational implicata it potentially induces, its morphology, its style or register, or its phonetic realization. Metalinguistic negation focuses, not on the truth or falsity of a proposition, but on the assertability of an utterance. (Horn, 1989, p. 363)

The problems discussed with reference to examples such as (2) and (4) have been closely related to the polemical notion that negation may be ambiguous. This was first introduced by Russell (1905/1988) as a case of semantic ambiguity. This view has been contested by authors who have specified that negation is not semantically ambiguous (as in Russell 1905/1988) but pragmatically ambiguous (Horn, 1989), vague (Atlas, 1977), underdetermined (see Leinfeller, 1994), or not ambiguous at all (Carston, 1996; Iwata, 1998).[2] What these views are trying to account for is the difference between answers like (5)b and (5)c to (5)a.

(5) a. We didn't see the hippopotamuses.
     b. We saw the rhinoceroses.
     c. We saw the hippopotami. (Carston, 1996, p. 310)

Carston (1996) argues that (5)b operates on the propositional content of (5)a by negating its truth value and is in some way descriptive of a state of affairs in the world, that "what we saw was rhinoceroses." Its propositional structure is *not P; Q*. However, (5)c does not have such a descriptive function, and a literal interpretation of it would lead to contradiction, that is, *P; not P*. Its function coincides with that described in Horn's quotation regarding metalinguistic negation, that is, an objection to an utterance on whatever grounds. The distinction between the descriptive and metalinguistic functions of negation was already pointed out by Ducrot (1972, p. 38), who argues that even though a metalinguistic negative utterance denies the validity of a previous proposition, a descriptive negative utterance does not. The contrasts between (6)a and (6)b illustrate this distinction.

(6) a. There are no clouds in the sky.
     b. This wall is not white.

Whereas (6)a illustrates the descriptive use of negation in Ducrot, that is, an utterance which describes a state of affairs; (6)b illustrates the metalinguistic use of negation, which will not be used descriptively of the entity *wall* but in order to deny a previous assertion about the wall being white.

Carston (1996, p. 310) provides further examples of metalinguistic negation, reproduced under (7).

(7) a. We don't eat tom[a:teuz] here, we eat tom[eiDeuz].
     b. He isn't neurotic OR paranoid, he's both.
     c. I haven't DEPRIVED you of my lecture on negation; I've SPARED you it.
     d. The President of New Zealand ISn't foolish; there IS no President of New Zealand.

The utterances under (7) are said to express an objection to a previous utterance, in (7)a, on phonologic grounds; in (7)b, with relation to the exclusion of a predicate said of an entity; in (7)c, on the choice of a lexical item; and in (7)d, we have the classical example of the negation of the presuppositions in a proposition. This is not the place to go into a detailed discussion of this topic, but it is important to point out, following Ducrot (1972, p. 38) and Carston (1996, p. 312), that negations of this type are not representations of states of affairs in the world, but, rather, representations of representations. Following this line of thought, Carston (1996) and Iwata (1998) defend an approach to metalinguistic negation as echoic in the sense used by Sperber and Wilson (1986, pp. 238–339). According to Carston, echoic uses are defined as follows: "A representation is used echoically when it reports what someone else has said or thought and expresses an attitude to it." (Carston, 1996, p. 320). A typical example is an ironic utterance, such as (8).

(8) The obnoxious beady-eyed woman is my wife.

Example (8) could be taken literally, but, most probably, it would be interpreted as being a repetition, an echo of what somebody else has said, thus yielding an echoic utterance with an ironic value. According to Carston (1996, pp. 320–321) its echoic nature is the crucial property of metalinguistic negation.

2.2.1.4. *Negation and generative semantics*
To end this introductory section on the traditional problems associated to negation, we mention those cases that have been of particular interest to generative semanticists (see Bolinger, 1977; Jackendoff, 1983) because they have brought up problematic issues with regard to the contention that variations in surface structure do not reflect variations in meaning. A classical example is that of ambiguity arising with regard to the scope of negation when quantifiers are present, as in (9):

(9) a. All the arrows did not hit the target.
     b. Many arrows didn't hit the target.

The two sentences under (9) cannot be said to be equivalent, since (9)a is ambiguous. The first interpretation is that none of the arrows hit the target, and the second interpretation is that some of the arrows (= not all) hit the target. Only in this last interpretation can it be considered to be equivalent to (9)b.

Finally, another much discussed aspect of negation has been that of negative transport or negative hopping, whereby the negative that should modify a verb in an embedded clause moves to modify the verb in the matrix clause, as in (10).

(10) a. I think he won't come.
      b. I don't think he'll come.

Processes of this kind, like the ones under (9), have been particularly interesting for generative semanticists (Bolinger, 1977), who have been concerned with the possible changes in meaning from one structure to another. Horn (1989, p. 343) proposes a reading of raising or negative transport as an indirect speech act, whose meaning is recovered by means of conversational implicature. In the case of some verbs, like *think, believe, want*, the meaning has become conventionalized, thus yielding short circuited implicatures.

Although the issues discussed in the preceding sections are not the direct concern of this book, they briefly illustrate what has been the state of the art with regard to negation for many years. In the following sections, the view of negation from the perspective of propositional logic is discussed.

2.2.2.   Negation in Logic

It is generally known that negation in natural language has different properties from those of negation in logic. It is necessary, however, to consider the characteristics of negation as a logical operator for two reasons: (1) because the properties of logical operators are also present in natural language; and (2) to establish the points where natural language differs from logic. This section is an introduction to the properties of negation as a logical operator both in simple structures and in complex structures such as contradiction and disjunction.

In logic, negation has the status of an operator that forms a compound sentence with a truth value that is the opposite of the truth value of the sentence it operates on (Allwood, Andersson, & Östen, 1974, p. 30). We can establish a table of correspondences (see Table 2.1) so that for every proposition P that is true there is another proposition $-P$ that is false, and vice versa (see also McCawley, 1981, p. 62).[3]

In linguistic terms, if the proposition *It's raining* is true, then the negative proposition *It's not raining* is false.[4] Negation can also be described in terms of set theory (Allwood, Andersson, & Östen, 1974, p. 31), so that a set A is defined as the set of all possible worlds where P is true. $-P$ will be the set of all possible worlds where P is false, which is the same as C A, or the complement of A (from Allwood, Andersson, & Östen, 1974, p. 31). The view of negation in terms of possible worlds or set theory can be illustrated informally by means of an example from *Catch-22*. There is an episode where the narrator is describing how people die in the war, outside the boundaries of the hospital. Negative statements are used to state *what is not the case* within the boundaries of the hospital.[5]

(11)   There was none of that crude, ugly ostentation about dying that was so common outside the hospital. They did not blow up in mid-air like Kraft or the dead man in Yossarian's tent, or freeze to death in the blazing summertime the way

TABLE 2.1.

P	$-P$
T	F
F	T

Snowden had frozen to death after spilling his secret to Yossarian in the back of the plane.

In terms of possible worlds, the description in (11) shows two mutually exclusive sets of states of affairs; as such, they describe complementary domains.

(12) a. People die violently. P
    b. People do not die violently. $-P$

This can be paraphrased as, for all the cases where P (People die violently) is true, $-P$ (People do not die violently) will be false, and vice versa. The domain where P applies is the domain indicated by the war, life outside the hospital, whereas the domain where $-P$ applies is the domain of the hospital.

In natural language, negation understood in broad terms can be expressed in different ways, such as the structures in (13) (Allwood, Andersson, & Östen, 1974).

(13) a. It is false that
    b. It is not the case that
    c. Not
    d. It is incorrect that
    e. It is not true that
    f. It is wrong that

The structures in (13a, b, d, e, and f) are paraphrases of external negation, however, (13c), represents the negative operator. The former rejects the truth value of a whole proposition together with its presuppositions. VP negation only negates part of a proposition, typically the predicate or a single constituent; in this case, the presuppositions are kept intact. Although this issue is controversial, as pointed out in Section 2.2.1.1., there is a tendency to argue that negation in natural language is fundamentally of the internal type, whereas the external type is more unusual in actual language use (Givón, 1979, p. 113; Lyons, 1981, p. 129).[6] This leads to one of the main differences between negation in logic and in natural language, pointed out by Allwood et al. (1974, pp. 29–32): (1) whereas in propositional logic, negation operates on a whole proposition, in natural language, almost any constituent can carry the negative operator, as in example (14).

(14) Nonstudents are not allowed.
    (Allwood et al., 1974, p. 31)

The negation of a NP—or of any constituent below the clause—is not possible in propositional logic. There are further differences between negation in natural language and negation in traditional logic:[7] (2) natural language has phonetic resources such as stress and focus to indicate constituent negation in a way that cannot be captured by traditional logic as in example (15).

(15) a.   Jack didn't hit Jill.
    b.   *Jack* didn't hit Jill.

(3)   Finally, and perhaps most importantly, logic cannot account for functional differences of negation in discourse. As Lyons (1981, p. 134) points out, logic cannot express the difference between assertion of negation and denial, as in (16), where (16)a is a negative assertion or negative statement, and (16)b is a denial.[8]

(16) a.   I noticed there was nobody in the room.
    b.   A: You shouldn't have bought that car.
        B: I didn't buy it, they gave it to me.

From the points discussed previously, it becomes clear that negation in natural language needs to be considered from a broader perspective than that offered by propositional logic. Contextual or pragmatic factors are crucial in determining the meaning and functions of negation in discourse, as is discussed in the following sections.

2.2.2.1. Negation in complex propositional structures: contradictions and disjunctions

Contradictions are complex propositions that are assigned the truth value *F* irrespective of the truth of the simple propositions. Thus, (16) is a contradiction.

(16)   It's raining and it's not raining.

This is represented in Table 2.2.

Although this is the structure that is normally understood to yield a contradiction, Escandell (1990, p. 924–925) draws our attention to the fact that other structures can also yield contradictory meanings. Thus, Escandell distinguishes between *formal contradiction*, which is in the structure above, and *nonformal contradiction*. Both are illustrated in examples (17).

(17) a.   It's raining and it's not raining.
    b.   It was good and it was bad.

TABLE 2.2.

P	$-P$	P and $-P$
T	F	F
F	T	F

Whereas (17)a illustrates the standard type of formal contradiction, (17)b is a form of nonformal contradiction. This distinction illustrates the classical distinction between *contradiction* and *contrariety* (see Horn, 1989, pp. 38–39). An opposition between contradictories involves the use of syntactic negation, as in (17)a and has the logical form p and *not-p*, however an opposition between contraries involves the opposition between predicates, as in (17)b and has the logical form p and q, where q equal *not-p* (Escandell, 1990, p. 924).

Escandell (1990, p. 925) further distinguishes between *simple* and *complex* forms of contradiction, depending on the structure of the proposition. This distinction is illustrated in (18).

(18) a. Her husband is not her husband.
     b. This bachelor is married.
     (Escandell 1990, p. 925)

Whereas (18)a is a complex formal contradiction, (18)b is a simple nonformal contradiction.

Contradictions, like tautologies, which are always assigned the truth-value T, have standardly been regarded as meaningless and uninformative in traditional semantic theory (see Leech, 1974, p. 75; Levinson, 1983, p. 194). Pragmatics has made an attempt to provide the means of interpreting contradictions as meaningful by making reference to their context of use. The claim that contradictions can be meaningful in a context is based on the intuition that expressions like (19) must have a meaning because they are used in everyday language.

(19) He's here and he's not here.

Escandell observed (1990, pp. 928–929) that there have been three different lines in the possible pragmatic explanation of utterances such as those in (19). The first possibility is to interpret them as being informative by means of conversational implicature. In this view, (19)a can be assigned a meaning that is recovered through inferencing procedures; this meaning may be that somebody is physically present but whose mind is on something else, different from the context where he is physically situated. The second possibility is to interpret them as special kinds of speech acts that are self-defeating, as in (20).

(20) I promise not to keep my promise.

As Escandell (1990, p. 929) observes, this view does not account for the fact that here the incompatibility lies in the relation between propositional content and illocutionary force, as the conditions for the production of the illocutionary act are not satisfied. In contradictions, this incompatibility does not arise because the sincerity conditions are satisfied and the speaker believes in the simultaneous validity of the two terms of the contradiction. This fact also invalidates the last pragmatic approach to contradiction mentioned by Escandell (1999, p. 930), Relevance theory (see Sperber & Wilson, 1986, p. 115). In this view, a contradiction is interpreted by eliminating one of the two contradictory terms.

> We assume that in these situations the contradiction is resolved by other means: for example, by a conscious search for further evidence for or against one of the contradictory assumptions. This seems to correspond to the introspective evidence that some contradictions are resolved by an apparently immediate and automatic rejection of the faulty premises, while other contradictions require deliberation. (Sperber and Wilson 1986, p. 115)

Sperber and Wilson's proposal thus seems to suggest that in general terms contradictions are not, strictly speaking, contradictory, as one term will always be favored against the other. Escandell (1990, p. 931) argues for a semantic-structural approach to contradictions where the interpretation of this kind of utterances does not depend on contextual information but can be recovered from the structure itself. This view hinges on a distributive interpretation of complex formal contradictions, as in (21) and (22).

(21) a.   It's raining and it's not raining.
     b.   In a sense it's raining, in another sense it's not.

(22) a.   I liked it and I didn't like it.
     b.   In a sense I liked it, in another sense, I didn't.

These examples can be interpreted by understanding that each of the terms in the contradiction is valid in a different domain (spatial, temporal, or other), indicated by the two conjoins in the coordination structure. A similar solution is proposed by Fauconnier (1985) in his model of mental spaces; in this theory, contradictions are understood as meaningful because the contradictory terms are elements in different cognitive domains, termed mental spaces, which are connected by a relation of accessibility. Although the approaches discussed so far may account for some types of contradictions, there is still a reluctance to accept the possibility of pure contradictions, that is, contradictions where the two terms must be simultaneously accepted as valid, within one single domain of reference. Contradiction is, however, a natural process in everyday language. To this respect, Givón (1984, p. 321) draws our attention to the fact that although contradiction is avoided by human beings as rational thinkers, it is used for other reasons.

> Unlike formal systems, humans are capable of compartmentalization, whereby contradictory beliefs held at the same time are rigidly segregated in subparts of the cognitive system, under different *personae*, etc. Further, humans are also capable of *change* or *faulty memory*, whereby they can hold contradictory beliefs in temporal

succession. Finally, they are also capable of *contextualizing* parts of their entire belief system, thus making the truth of some propositions vary with the change of internal or external context. (Givón, 1984, p. 321)

Thus, contradiction can arise in language use because of the internal inconsistencies or complexities of a person's belief system, or because of the effects of change through time or in different contexts with regard to beliefs previously held. This implies a reformulation of the logical notion of contradiction as a more flexible discourse phenomenon that involves the denial of the truth of a previously held proposition within a domain where such a denial would not be expected.[9] Such a domain is, for example, a person's attitude to a particular issue. A further reformulation is necessary to account for contradiction in discourse when the inconsistency that arises cannot be solved, precisely because the inconsistency is deliberate. Example (58) discussed in Section 2.4. illustrates this last type of contradiction.

In natural language, then, an operation such as contradiction can convey a meaning that cannot be expressed by a strictly logical approach. As Givón (1989, p. 167) observes, whereas in logic we have the two extremes of tautology (propositions which are always true) and contradiction (propositions which are always false), natural language is a hybrid system, "a compromise between the two extremes."

   (a) **Tautologies:** total informational redundancy; no interest.
   (b) **Contradictions:** total informational incompatibility; no coherence.
   (Givón 1989, p. 268)

Thus, most propositions in natural language fall between these two extremes because they are "informational hybrids which carry some presupposed (old) and some asserted (new) information" (Givón, 1989, p. 167).

Disjunction is another complex proposition type that corresponds to the function of *or* in natural language. It can be *exclusive*, in which case the proposition will be true if and only if both conjoins are true, or *inclusive*, which will be true if only one of the conjoins is true. Allwood, Andersson, & Östen, (1974, p. 34) provide the following examples of true and false disjunctions.

   (23) a. Mars is a planet or a black hole. (T)
        b. Mars is a satellite or a black hole. (F)

In terms of set theory, "the truth-set for $p \lor q$ will be the set of all worlds where *p or q* is true—which is the same as the union of *A* and *B*." (Allwood et al., 1974). Like contradiction, disjunction in natural language functions differently from the way it does in logic, as the latter does not capture the uncertainty typical of the *either–or* structure in common language use. Allwood and colleagues (1974, p. 36) point out that in logic it is possible to build a structure like that in (24) while looking out of the window.

(24)   Either it's raining or it's not raining outside.

In natural language, an utterance such as (24), although possible, would be odd, as disjuction is usually associated with the expression of doubt or potentiality of some kind, as in (25).

(25) Either he's at home or he's left (but I don't know which).

Furthermore, in natural language, disjunction is very often of the *inclusive* type, whereas in logic, disjunction is generally understood to be *exclusive*.[10] This distinction has consequences in language use that cannot be captured by logic. In logic, disjunction establishes a binary contrast between two mutually exclusive terms that together form a set. Whether the truth value assigned to it is T or F has no consequences on a given context of use. In natural language use, however, disjunction has consequences that stem from the cognitive organization of experience into mutually excluding options. As pointed out by psychologists (see Apter, 1982; Clark & Clark, 1977) and linguists (Grice, 1975; Lyons, 1977), there is a tendency to categorize experience in terms of binary systems of opposite terms, to talk of things as either black or white, good or bad, and so forth. In many cases, the binary contrast admits a middle term, an alternative to the terms referred to in the disjunction (*gray, neither good nor bad*). This middle term (or term from a different set altogether) is often overlooked, however, sometimes with dramatic effects on the way experience is conceptualized.[11] We come back to this point in further sections throughout this chapter and the next.

2.2.2.2.   *The Laws of Non-Contradiction and of the Excluded Middle*
In logical terms, both contradiction and disjunction are closely related to what in propositional logic is referred to as the law of non-contradiction and the law of the excluded middle (see Horn, 1989, p. 79). The Law of Non-Contradiction establishes that if a proposition P is true, *not-P* is false (see Givón, 1993, p. 187; Horn, 1989, p. 18). The Law of the Excluded Middle establishes that a proposition P is *either* true *or* false, thus excluding the possibility of having a middle term. Both laws exclude the possibility of being and nonbeing at the same time (Horn, 1989). As was pointed out, however, natural language functions in ways where the Laws of Non-Contradiction (LC) and of the Excluded Middle (LEM) are not followed.[12]

In this respect, it is interesting to observe that the acceptance of the LC and the LEM is exclusively a Western phenomenon, as the Eastern philosophical tradition does not consider them as laws of logic (see Horn, 1989, pp. 79–96). Thus, Horn (1989, pp. 79–80) points out that by means of the *principle of four cornered negation*, contradiction and excluded middle are parts of a process where no final solution is provided. The principle states the propositions that would "describe a subject S in relation to an entity or class P" (1989), as presented in (26).

(26) a. S is P.
     b. S is not P.
     c. S is both P and not-p.
     d. S is neither P nor not-p.

As can be observed from example (26), the process represented here infringes the LC and LEM followed by Western logic because it involves the simultaneous acceptance of contradictory propositions. In Eastern logic, there is a long tradition of meditation as a means of achieving an insight as a solution to a logical problem, rather than the rational process typical of Western thought. Consequently, the task of the would-be philosopher or "sage" would be to consider in turn each of the propositions in (26) and reject them all (Horn, 1989, p. 80). Although the discussion of such philosophical differences is not the direct concern of the present book, it is interesting to observe that, although in Western tradition the transgression of certain logic principles will indicate inconsistency, falsity, or incoherence, the fact is that this is only one view. In other traditions, such as the one mentioned, anomalies such as contradiction are not considered to be anomalous at all, but rather as one more element in the process of perception and observation of reality.[13] I come back to this point in Chapters 4 and 5, where examples of different types of contradictions are discussed.

2.2.3. Negation in Psychology

Negation has been the focus of attention of many studies in psychology, both because of its cognitive properties as contrasted with the expression of the affirmative, and because of its significance in therapeutic processes. From a linguistic perspective, both aspects are significant, because the former provides insights regarding the ideational component of negation, whereas the latter does so with regard to interpersonal aspects of negation.

The findings on the cognitive processing and production of negatives have been very influential on the understanding of negation as a linguistic phenomenon. In particular, the cognitive properties of negation constitute the main evidence in favor of considering it the marked term in the polarity system. It is generally agreed that negation as a structure involves the formation of a complex structure with regard to the corresponding affirmative, and that this complexity is evident in the different perspectives from which negation has been studied.

From the point of view of language acquisition, negative structures are acquired later than affirmative structures, that is, the negative particles are incorporated into previously learned affirmative structures; however, the process of acquisition and production of negation in general terms is rather more complex, as children learn quite early to say "no," or to express negation (nonexistence, refusal, and rejection) by nonlinguistic means (see Clark & Clark, 1977, pp. 348–351).[14] This process has a parallel in diachronic change in language because negation is incorporated into many languages once the affirmative structure has already been developed (Givón, 1979, p. 121). Furthermore, as compared to the processing of affirmative structures, negatives take longer to process. This has been proved by different studies carried out by means of making subjects go through tests directed at the identification of the processing time of affirmative and negative sentences (Clark, 1976; Just & Clark, 1973; Wason, 1965). These results are also applicable to inherently negative lexical items, such as *absent*, which are harder to process than their positive counterparts, such as *present* (see Clark, 1976, p. 42).

Similarly, other tests have concentrated on the identification of factors that condition the production of negative sentences. In this case, the pioneer was Wason (1965), who established that the production of negative structures takes place when there is an expectation that is being defeated (Wason, 1965, p. 7).

These findings seem to indicate that there is an asymmetrical relation between negative and affirmative structures, the negative being a second-degree operation that is realized on a preexisting affirmative structure or proposition.[15] This view is obviously present in approaches to negation based on the idea of the "denial of an expectation," which is discussed in Section 2.5.4.[16]

A crucial issue in the psychology of negation is its role in the perceptual coding of experience. To this respect, there are several observations to be made.

1. Experience is usually coded in positive terms, rather than in negative terms. Clark and Clark (1977, p. 240–241) point out that tests on subjects asked to make descriptions of places invariably yielded the use of the affirmative. Denials will be used, however, when the speaker wants to deny an expectation he or she assumes that the listener is holding (see Clark & Clark, 1977, p. 99; Wason, 1965, p. 7).

2. With regard to negation in the lexicon, there seems to be a close relation between the use of positive terms to express extent and the use of negative terms to express lack of extent (Clark & Clark, 1977, p. 538), a phenomenon which is apparently universal; thus, the terms *long, tall, wide* are considered to be positive against their corresponding opposites *short, low, narrow*, which are classified as negative.

3. Finally, positive terms are more informative than negative terms; we describe things by what they are and not usually by what they are not (see, for example, Clark, 1976, p. 54). Thus, in pointing to my house, I will normally say (27)a and not (27)b.

   (27) a. That is my house.
       b. That is not my parents' house, that is not my friend Gina's house, that is not my teacher's house, and so forth.

These arguments provide evidence in favor of the fact that experience is usually coded positively, and that negation tends to be used only if there is an expectation that is not fulfilled. These characteristics are expanded from a linguistic perspective in Sections 2.5.4., 2.6.1. and 2.6.4., where the marked character of negation and its presuppositional nature is discussed.

With regard to the second aspect of negation mentioned at the beginning of this section, negation is also relevant as a stage in a therapeutic process, or, in psychoanalytic terms, the use of negatives is significant as a means of revealing character traits and the emergence of subconscious material into consciousness (see Freud, 1976a; Labov & Fanshel, 1977). These observations are relevant to linguistic research because language use reveals information about the people and communities who use it. If we are dealing with fiction, as in the case of the present book, information about the psychological features associated with specific language uses can provide insights regarding the characters who use language in a particular way.

With regard to the significance of negation in a psychoanalytical perspective, there are two observations to be made.

1. Negation is standardly used by patients during therapeutic processes to deny information from the subconscious, which has been repressed. Negating this material is the first step toward the achievement of an awareness of such material, which is usually traumatic (see Freud, 1976a, p. 253; Labov & Fanshel, 1977, pp. 334–335). This phenomenon is relevant to linguistic research because it is obviously based on the idea that negation both negates and affirms at the same time. Though this is not, strictly speaking, true, I show in further sections that negation in certain specific uses actually works in this way, more precisely in the phenomenon called accommodation.

2. The second observation to be made in terms of the relevance of negation from a psychoanalytic perspective is the well-known fact that there is a close relation between negation and acts of resistance, such as the one mentioned previously in a therapeutic process, and other challenging acts, such as rejection or refusal. A consistent use of acts of this type might be indicative of rebelliousness of some kind, and illustrate conflictive behavior on the part of characters in fiction. This is illustrated in Chapter 5, where the functions of negation in *Catch-22* are discussed.

2.2.3.1.   *Contradiction and paradox as cognitive synergies*

In the sections under 2.2.2., the issue was introduced that, although the organization in terms of opposites is widely accepted as a universal tendency in language, the simultaneous presence of opposite terms to yield contradictory statements or ambiguities is generally regarded as anomalous, in the sense that phenomena of those types are logically unacceptable, linguistically deviant, and psychologically disturbing. Apter's (1982) theory of psychological reversals provides an attractive alternative to more traditional views, which focus on normative and unproblematic behavior. In this section we discuss some aspects of Apter's (1982) theory that are relevant to the interpretation of contradiction as a phenomenon that involves negation in *Catch-22*.

Apter (1982, p. 8) establishes that the theory presented is concerned with inconsistency and paradox as part of human behavior and motivation. According to this theory, it can be said that inconsistency pervades all human behavior, in the sense that as individuals, we are subject to variation with respect to different factors. Thus, two individuals may react differently to the same situation, and, furthermore, the same individual may react in different ways to the same situation at different points in time.

The central notion in Apter's model is the notion of cognitive synergy, which has to do with the way in which we perceive contradictory meanings as coexisting. The phenomenon is described as follows by the author.

> The idea is that opposite characteristics may co-exist in the sense that one is aware of both in consciousness, in relation to a given identity, and that these opposites both contribute something to the full meaning of the identity, or contribute alternative meanings to the identity. Either way, synergies always embody some form of self-contradiction. (Apter, 1982, p. 141).

The author gives the simple example of the checker-board, which is constituted by mutually exclusive attributes, black squares and white squares; however, as a whole, it can be said to be both black and white. The notion of cognitive synergy develops from the observation that, not only do we manifest a tendency to interpret experience in terms of opposites, but also that in certain cases, one identity can be assigned two opposite characteristics and display contradictory meanings. As the author points out (1982, p. 141) "Although this may not be logically possible, phenomenologically it is prevalent." Furthermore, a synergy has the peculiarity of producing an effect that could not be achieved by the two constituting elements in isolation. The term *synergy* is used in other disciplines, such as medicine, to indicate this meaning, such as when it is said that a mixture of alcohol and certain drugs may have unpredictable effects.

Apter (1982) provides further examples to indicate the different ways in which contradictory meanings can coexist. For example, if one is not sure whether a person is male or female, he or she will try to find evidence in favor of one of the two interpretations by identifying relevant properties in a process where opposite properties will be assigned in turn. In this sense, the process is described as bi-stable, as properties belonging to each of the two domains are focused on at different moments.[17]

We can take an example from *Catch-22* to illustrate how this kind of process is exploited in the novel *Catch-22*. In example (28), we have a comment on a character who is a friend of the protagonist. He is described as follows.

(28) McWatt was the craziest combat man of them all probably, because he was perfectly sane and did not mind the war. (p. 80).

In McWatt's description we have the definition of an identity (McWatt) by means of two opposite terms (*sane* and *crazy*), defined by Apter (1982) as *strong opposites* (or contraries, in Horn's terminology), as opposed to complementaries such as *sane–not sane*, which he defines as *weak opposites* (or contradictories). It can be said that the simultaneous perception of both attributes gives place to a cognitive synergy, which requires the acceptance of both terms. The triggering of knowledge of the world, and of wars in particular, including reactions to being part of a war, makes the reader try to solve the contradiction by interpreting the terms as belonging to different domains. Thus, McWatt is crazy from a subjective perspective because he does not manifest an expected reaction (for the narrator in the novel and many readers), that is, one of worry, despair, and so forth. In this sense, he is crazy because his "anomalous" reaction is implicitly contrasted to the more usual and, therefore, sane reaction. Furthermore, he is sane from an objective perspective, that is, he is not clinically crazy with respect to ordinary behavior, except for what has been described. This leads to a perception of the character as being both crazy and sane, each of the properties being focused on depending on the domains of reference. This kind of contradiction illustrates one of the types discussed by Escandell (1990) and mentioned in Section 2.2.2.1.

It is interesting to observe that contradiction and ambiguity can be perceived as either threatening or as a stimulating and enjoyable phenomena depending on the mental state of the receiver (Apter, 1982, pp. 145–153). Also, contradiction is typical of certain domains of human activity, such as art, as discussed in Chapter 1. This view may account for the different reactions to literary works such as *Catch-22*, which can either be rejected as nonsensical and incoherent or enjoyed because such inconsistencies can be perceived as funny and challenging in a positive way. Chapter 4 in this book focuses on the relation between contradiction and humor.

To sum up, it becomes clear that phenomena in which negation is involved, such as contradiction, may be logically unacceptable but from a psychological point of view, they are not only justified but they are also meaningful. Determining what kind of meanings may emerge from contradictions has traditionally been very problematic, both in psychology and in linguistics. Apter's (1982) notion of cognitive synergy has been proposed as an alternative to traditional theories, which consider contradiction as marginal and anomalous. In further chapters of the book, I account for this phenomenon from a linguistic perspective.

2.3. DESCRIPTIVE APPROACHES TO NEGATION

In the sections that follow, the notion of negation as a linguistic concept by considering the classifications that are standardly made of negative words is introduced.

2.3.1. A Classification of Negative Words: Explicit and Implicit Negatives

As has already been discussed, negation understood in broad semantic terms can be expressed in different ways in natural language. In this section, the main distinctions followed currently in the literature is described and a classification of negative words according to morphosyntactic and semantic criteria is established.

The main problem involved in the identification and classification of negative words has been the lack of correspondence between word content and word form, already observed by Jespersen (1917/1966, p. 22). Not only is it the case that there are words with no overt mark of negation (*absent, fail, lack, forget*) which, however, are generally understood to convey a negative meaning (Huddleston, 1984, p. 428; Jespersen, 1917/1966, p. 38; Quirk, Greenbaum, & Svartvik, 1985), but also, there are cases where there is a lack of fit between the grammatical structure of an utterance and its force (see Tottie, 1991, p. 34). In the latter case, we may well have negative utterances with the force of agreements, as in (29), or, conversely, affirmative utterances with the force of refusals, as in (30).

> (29) a.   . . . and I didn't like his attitude at all.
>   c.   No . . .

> (30) a.   Look, you just have to leave.
>   b.   I'm staying here.

This section focuses on a description of negative words; the discussion of negation and force is dealt with in Sections under 2.4.

Klima (1964) was the first to attempt to establish a formal distinction between words that could be identified as negative both in form and meaning and words that are negative in meaning but not in form.[18] Since then, the tests of co-occurrence of negative words with nonassertive terms, such as *any* and *either* in coordinated structures and the combination with positive tags, have been standardly applied to identify what have been called *explicit negatives*. This can be observed in (31) and (32).

> (31) a.   He never told us in time, and she didn't either.
>   b.   *He never told us in time, and she did too.

(32) a. He never told us in time, did he?
     b. *He never told us in time, didn't he?

By explicit negatives, the following group of negative words is understood: *not, n't, no, nobody, no one, nowhere, nothing*. These words are negative in meaning, they are marked morphologically for negation and they follow co-occurrence restrictions that single them out as syntactically negative. They are referred to by different authors as *clausal negation* (Huddleston, 1984, p. 423; Quirk et al., 1985, pp. 777–782; Quirk & Greenbaum, 1990, p. 223), *syntactic negation* (Givón, 1993, p. 202) or *nuclear negatives* (Downing & Locke, 1992, p. 180). Huddleston (1984, pp. 423–424) proposes further tests to distinguish what he calls *clausal* negation from *subclausal* (or constituent) negation, such as fronting, which in the case of clausal negation forces subject–verb inversion; subclausal negation does not.

Syntactic negation usually includes also the group of *broad* negatives or seminegative words formed by the adjuncts *hardly, scarcely, seldom, rarely*, and the determiners *few* and *little*. Although these words have negative meaning, they have no morphologic indication of a negative affix or particle, unlike the negatives mentioned previously (see, for example, Quirk et al., 1985, p. 780). Because co-occurrence tests show that they tend to function like explicit negatives, however, they are usually classified in this group. Thus, they can combine with *at all* (see Jespersen, 1917/1966, p. 38; McCawley, 1995, p. 32), an expression identified as co-occurring with nonassertive forms only, and with the nonassertive forms *any* and *ever*, although they do not combine with affirmative tags. This can be seen in examples (33) to (36).

(33) a. ? He hardly recognized her at all.

(34) a. *He hardly recognized her, did he?
     b. He hardly recognized her, didn't he?

(35) a. He hardly ever discussed any of his problems.
     b. *He hardly never discussed some of his problems.

(36) a. He had hardly had any breakfast.
     b. *He had hardly had some breakfast.

Further, distinctions are established between negation of the verb and negation of other constituents (henceforth, V-neg and constituent-neg respectively). Thus, Jespersen (1917/1966, p. 42, and 1924a, p. 438) distinguishes between "nexal" (V-neg) and "special" negation (constituent-neg and lexical neg.). Quirk, Greenbaum, & Svartvik (1985, p. 775) distinguish between what they call *clausal negation* and *local negation*. Huddleston (1984, pp. 419–420) distinguishes between *clausal* and *subclausal negation*. Givón (1993, p. 202) distinguishes between *syntactic, morphological* (words with negative affixes) and *inherent* negation (words assigned a negative value in a pair of opposites). Tottie (1991, p. 9) talks about *not-negation* and *no-negation*, respectively.

As Clark (1976, p. 33) points out, Klima's (1964) tests leave out what are usually referred to as implicit negatives, or words that convey a negative meaning although their syntactic co-occurrence rules are the same as for positive terms, that is, they combine with assertive terms and with negative tags, as shown by the examples under (37).

(37) a.   John is unhappy, and Mary too.
     b.   *John is unhappy, and Mary either.
     c.   John is unhappy, isn't he?
     d.   *John is unhappy, is he?

Implicit negatives is a term used to refer both to words where the negative meaning is indicated by affixation, or "morphological" negation (Givón, 1993, p. 202), and words which are inherently negative, or "inherent negation."

Morphological negation can be indicated by prefixes, such as *in-, im-, il-, dis-, un-* in English, or by the suffixes *-less*, and *-out*, as in *without*.[19] Even the identification of morphologically marked negative terms might turn out to be problematic, however, because not all affixes indicate the same type of relationship between positive and negative terms. Thus, although there are clear-cut cases, such as those in (38), where the negative term indicates an opposite value to that expressed by the positive term, the examples in (39) express a different kind of relationship, and those in (40) lack a corresponding positive term.

(38)   true–untrue, expected–unexpected, legal–illegal

(39)   appear–disappear, load–unload

(40)   uncouth, restless, disgruntled

In the case of inherent negatives, the identification of such terms is still more problematic, both for practical and theoretical reasons, as is pointed out by several authors, from Jespersen (1917/1966, p. 43) to Tottie (1991, p. 7). In practical terms, the task of identifying inherent negatives in a corpus is practically unthinkable (Tottie, 1991, p. 7). Furthermore, from a theoretical standpoint, nothing prevents us from reversing the process by which we assign a negative value to a word (Jespersen, 1917/1966, p. 43). That is, although we usually think of a word like *fail* as meaning *not succeed*, we may just as well think of *succeed* as meaning *not fail*.

In brief, a distinction is established in general terms between words that are *explicitly* negative, or syntactic negation, and words that are *implicitly* negative, or morphological and inherent negation.[20] I refer to the latter also as *lexical negation*, as it involves the expression of negation through lexical rather than syntactic means.

2.3.2.   Syntactic Negation Types

Syntactic negation is typically carried out by means of negating either the lexical verb or the auxiliary in clause structure, as in (41).

> (41) a.   He didn't wake her up.
>   b.   Not wanting to wake her up, he tried to leave quietly.

Syntactic negation can also be carried out by negating nonverbal constituents (see Downing & Locke, 1992, p. 180; Givón, 1993, pp. 198–199; Huddleston, 1984, p. 419; Quirk, Greenbaum, & Svartvik, 1985, p. 790). The possibility of attaching the negative to almost any constituent was pointed out by Jespersen (1917/1966, p. 56, 1924/1961a, p. 446), who described the two opposing tendencies manifested in negative attraction: on the one hand, a "universal tendency to attract the negative to the verb even where it logically belongs to some other word" (Jespersen, 1917/1966, p. 56), and, on the other hand, another tendency to attach the negative particle to "any word that can easily be made negative" (Jesperson, 1917/1966, p. 56). The following is a classification based on Downing and Locke's work (1992, p. 180); it shows the different types of syntactic negatives, or *nuclear negatives*, where a constituent different from V is negated.

(42)			
Negative pronoun	at S		Nobody knows.
	at Od		I told nobody.
Negative specifier	NP det.		I felt no pain.
	NP predet.		Not many people left.
	AdjP. spec.		It's none the worse...
	Adv. spec.		...not even...
Negative adjunct			I've never been there.
Negative coordinator			Neither this nor that.

The choice of V-neg versus constituent-neg is governed by structural and pragmatic factors. This is mentioned by different authors (Downing & Locke, 1992, p. 181; Jespersen, 1917/1966, p. 56, 1924/1961a, p. 446) and is dealt with in depth by Tottie (1991) in a monograph on negation in English speech and writing. Here, we simply outline some of the relevant issues involved in the choice of the two forms of negation.[21] From a pragmatic perspective, stylistic and contextual factors condition the choice of the negative form: V-neg is preferred in spontaneous, informal speech; constituent-neg is preferred in formal, written varieties of English. There are, however, structural constraints on the choice of the negative form. Thus, if the negative occupies initial position, only constituent-neg is possible.[22] As is pointed out by Downing and Locke (1992, p. 181), this is closely connected with the phenomenon of the scope of negation, which is discussed as follows.

2.3.3.   The Scope of Negation

The scope of negation can be defined as the semantic influence that the negative item exercises over the constituents of the clause where it appears, or the semantic domain on which negation applies (see Downing & Locke, 1992, p. 182; Givón, 1993, pp. 197–198; Huddleston, 1984, pp. 428–432; Quirk, Greenbaum, & Svartvik, 1985, pp. 787–790). Usually, all the constituents that follow the negative fall under the scope of negation, while the subject remains outside. This can be observed by the fact that assertive forms can occupy subject position, while nonassertive forms will be found in other positions, as in example (43).

(43)   Some people don't have any sense of humour. (Downing & Locke, 1992, p. 182)

In (43), the nonassertive form *any* lies within the scope of negation. The subject pronoun *Some* is outside the scope of negation. Assertive forms can sometimes occupy positions following the verb carrying the negative, but in this case, the meaning is different from that expressed by a corresponding clause with a nonassertive form. Downing and Locke (1992, p. 182) provide the following examples.

(44)   a.   He didn't reply to any of my letters.
        b.   He didn't reply to some of my letters.

In (44)a, the scope of the negative extends to the end of the clause and yields the meaning that *None of the letters received a reply*, whereas (44)b yields the meaning that *Some of the letters received a reply*.

The scope of negation can be indicated by means of contrastive stress, which narrows down the scope of negation to the constituent that receives the focus, leaving the rest of the clause presupposed (Givón, 1993, p. 197; Quirk, Greenbaum, & Svartvik, 1985, p. 789). This can be observed in the following examples.

(45)   a.   John didn't hit Bill.
        b.   **John** didn't hit Bill.
        c.   John didn't hit **Bill.**

In (45)a, we have an example of neutral negative focus, which, consequently, involves the negation of the whole predicate. In (45)b, we have subject focus (someone hit Bill, but not John), and in (45)c, object focus (John hit someone, but not Bill). According to Huddleston (1984, pp. 432–434), contrastive stress selects the foregrounded entailment that is applicable in a given situation, thus specifying the scope of negation for that particular utterance.[23]

Adjuncts tend to attract the focus of negation, so that the rest of the clause is presupposed. Givón (1993, p. 197) argues that this accounts for the fact that in examples like (46)b and (46)c, only the adjunct is understood to fall under the scope of negation, and not the verb or any other constituent.

(46) a. She ran as fast as she could.
       b. She didn't run as fast as he could.
       c. She didn't write the book for her father.

In (46)b, the adjunct *as fast as she could* falls under the scope of negation, though the presupposition *she ran* is kept unchanged, the same as in the affirmative (46)a. The same process takes place in (46)c, where the adjunct *for her father* falls under the scope of negation, whereas the presupposition *she wrote the book* is kept. Givón (1993, p. 199) argues that the reason for optional constituents to attract the focus of negation may be the fact that these constituents are the focus of assertion, even when the structure is affirmative. Thus, Givón argues that the following pragmatic inference operates in such cases: "If an optional element is chosen, chances are it is the focus of asserted information." (Givón, 1993, p. 199).[24]

In cognitive approaches to negation (see Givón, 1979, 1984, 1993, and especially, Langacker, 1991), the notion of scope must be further understood as a complex conceptualization process that arises from the interaction of two predicates or structures. This view relies on what is known as the asymmetricalist presuppositional view of negation (mentioned in Sections under 2.2. and discussed in detail in the Sections under 2.5.).

2.3.4.   Lexical Negation

The issue of negative polarity in lexical items was introduced in Section 2.2.3. Here, the topic is again taken up in order to provide further considerations from a linguistic, rather than a psychological, perspective. The view defended here is still the one presented previously, however, which argues for a cognitive basis in the way in which polarity distinctions are manifested in the lexicon.

2.3.4.1.   The cognitive basis of binary oppositions in language

It is a natural tendency in human language to classify experience in terms of opposites, which are manifested linguistically as pairs of antonyms (see Apter, 1982, p. 137; Clark & Clark, 1977, p. 426; Cruse, 1986, Chapters 10, 11, and 12; Givón, 1984, p. 351; Horn, 1989, p. 39; Lyons, 1977, p. 271; Werth, 1984, p. 22). It was pointed out in Section 2.2.3. that, in terms that express perceptual coding, the term that indicates lack of extent is assigned a negative value, whereas the term that indicates extent is assigned a positive value. This has further consequences on language use because the positive term "neutralizes" and becomes the term that is used to indicate the whole scale that is being referred to. Negative terms are never used in this way (see Clark & Clark, 1977, p. 426). This can be observed in the examples under (47).

(47) a.   How long was the movie?
     b.   ? How short was the movie?

Whereas (47)a is the neutral way of asking about the length of a movie, (47)b would only be possible in specific contexts, for example, when the notion of "shortness" has already been introduced, by means of an utterance like (48).

(48)   The movie was very short.

Indeed, Lyons (1977, pp. 275–276) argues that utterances of the type illustrated in (47)b presuppose the negative property. Utterances like (47)a are neutral in this respect.

Although the apparent arbitrariness of the linguistic assignment of positive and negative values to lexical items has been observed on several occasions (see, for example, Jespersen, 1917/1966, p. 43), in psychology and cognitively based linguistic theories, it is argued that the linguistic system of coding such values is far from arbitrary (see Clark & Clark, 1977, pp. 534–535; Givón, 1984, p. 351; Langacker, 1991, pp. 132–139; Werth, 1995c, pp. 32–33). Rather, as discussed previously, the assignment of positive and negative values follows cognitive principles of perceptual saliency versus perceptual backgroundedness.

This can also be observed in terms that not express perceptual coding but an evaluation, such as *good–bad, mortal–immortal, legal–illegal*. In these cases, there is a natural tendency to distinguish between a positive and a negative term in these oppositions, even when there is no morphological mark of negation (as in *bad*). Clark and Clark (1977, p. 539) argue that the term that usually expresses the expected norm, the typical state, will be categorized as positive; however, a departure from a norm or the taken-for-granted status quo will be categorized as negative. In these terms, *bad, immortal,* and *unusual* indicate departures from more frequent—or desirable—norms indicating what is *good, mortal,* or *usual*. Hence "Normal states are conceived of positively, and abnormal states as the absence of normal states, as negative states" (Clark & Clark, 1977, p. 539).[25]

In an insightful discussion of the different types of relationships between opposites, Cruse (1986, p. 197) observes that opposites have a paradoxical nature, as they are simultaneously maximally separated and close. Their closeness has to do with the fact that opposites have the same distribution and frequency. Cruse (1989, p. 197) further argues:

The paradox of simultaneous difference and similarity is partly resolved by the fact that opposites typically differ along only one dimension of meaning: in respect of all other features they are identical, hence their semantic closeness; along the dimension of difference, they occupy opposing poles, hence the feeling of difference.

2.3.4.2. Contraries and contradictories

Contrary and *contradictory*[26] are the terms used by Horn (1989, p. 39), on the basis of traditional classifications dating from the time of Aristotle, to distinguish between opposite terms that are gradable (contraries) and those that are not gradable (contradictories). This distinction is expressed by other authors by means of the terms *antonym* and *complementary*, respectively (Cruse, 1986, pp. 197–222; Lyons, 1977, p. 279; Werth, 1984, p. 158, 1999, p. 128). In any case, a contradictory or complementary is a term that is governed both by the Law of Non-Contradiction and the Law of the Excluded Middle. A contrary is a term that is only governed by the Law of Non-Contradiction (Horn, 1989, pp. 270–271). This can be observed in the following examples, discussed by Werth (1984, p. 158).

(49) a.   John is neither clever nor stupid.
     b.   ? John is neither dead nor alive.

Example (49)a contains two contraries, *clever–stupid*, that is, two gradable opposites that exclude each other but accept the existence of middle terms between the two ends of the scale; example (49)b contains two contradictories, *dead–alive*, two terms that mutually exclude each other without accepting the possibility of a middle term.

It is interesting to observe that there is a tendency in human communication to use contraries as if they were contradictories, thus accentuating the binary opposition between the two extreme terms, rather than the choice between several items on a scale (see Grice, 1975; Horn, 1989, p. 332; Lyons, 1977, p. 278). Thus, Lyons (1977, p. 278) observes that "If we are asked *Is X a good chess-player?* and we reply *No*, we may well be held by the questioner to have committed ourselves implicitly to the proposition that X is a bad chess-player." From a logical standpoint, there is a lack of fit in the relation between two contraries, however, there is no inconsistency in the case of contradictories (Lyons, 1977, p. 272). Thus, the proposition in (50)a implies (50)b and (50)c implies (50) d, indicating relationships of logical consistency.

(50) a.   X is dead.
     b.   X is not alive.
     c.   X is not alive.
     d.   X is dead.

In the case of contraries this is not the case. Although (51)a implies (51)b, (52)a does not imply (52)b.

(51) a. X is not clever.
    b. X is stupid.

(52) a. X is not stupid.
    b. X is clever.

Contrariety can, furthermore, be applied to the different types of oppositions illustrated in (53).

(53) a. black/white
    b. black/red

In this case, we have oppositions of terms belonging to a multiple-member set, rather than the oppositions between members at the end of a scale, as in the examples above. Contraries like those in (53)a have been called *polar opposites*. Those in (53)b are also referred to as *disjuncts* (see Horn, 1989, p. 39).

I henceforth follow Horn's classification of opposites, which is organized as follows. *Contradictories* refer only to the contradictory status of a proposition, that is, this term will not be used as applied to lexical items, but only to propositions. Among contraries, the author distinguishes between the following types: *mediate contraries* or *weak contraries*, and *immediate contraries* or *complementaries* (odd/even). Among *weak contraries*, we can further distinguish between *disjuncts* (black/red) and *polar opposites* (black/white). This classification is reproduced in the following diagram.

(54) a. contradictories: black/not black
    b. contraries
       1. weak contraries
          a. disjuncts: black/red
          b. polar opposites: black/white
       2. strong contraries or complementaries: odd/even

To end this section, it has to be pointed out that there is a lack of fit between an item that is negated syntactically and the corresponding morphological and inherently negative forms (see Givón, 1984, p. 342; Horn, 1989, p. 334; Werth, 1984, p. 152). Example (55) illustrates the scale on which such a relation can be represented.

(55) a. He isn't happy.
    b. He is unhappy.
    c. He is sad.

From a logical point of view, the three forms represented in (55) should be equivalent; however, they are not. Givón (1984, p. 132) argues that there is a difference in degree of strength of the utterance, which varies depending on the criterion adopted: if the scale is measured in terms of speech act strength, the syntactic negative (55)a is the strongest and the inherent negative (55)c is the weakest. But in terms of subjective certainty, it is the other way round. This seems to be related to the fact that syntactic negation is felt to be more vague than lexical negation. Although from a speech act perspective, syntactic negation yields strong assertion, versus the weaker forms containing lexical negation.

2.4. APPLICATION TO EXTRACTS FROM *CATCH-22*

In this section some of the issues discussed in the present chapter and their applicability to the analysis of extracts from *Catch-22* are considered. Of particular interest, is the attempt to prove the following points: (1) negation cannot be accounted for only in terms of its definition as the reversal of the truth-value of a proposition; and (2) contradictory statements are meaningful, but their interpretation cannot be carried out in terms of propositional semantics. The rejection of an analysis of negation based on logic and traditional semantics points to the need of exploring further possibilities of analysis. Extract (56) illustrates the occurrence of syntactic negation, and extract (58), the occurrence of contradiction in discourse.

Let us first consider the example that contains syntactic negation.

(56) To Yossarian, the idea of pennants as prizes was absurd. No money went with them, no class privileges. Like Olympic medals and tennis trophies, all they signified was that the owner had done something of no benefit to anyone more capably than everyone else. (p. 95)

In (56), we have the following examples of syntactic negation, or, more precisely, constituent negation: (1) *No money went with them; (2) no class privileges [went with them]; and (3) the owner had done something of no benefit to anyone*. . . . An analysis based on propositional logic or on structural analysis would not be able to account for the functional properties of these sentences, an approach which is necessary if our aim is that of describing the function of negation in discourse.

The first problem that arises when analyzing a passage such as the one in (56) by using traditional logic is that it contains examples of constituent-negation, and not of V-negation or the negative logical operator. The negative sentences in (56) can be said to be equivalent in meaning to propositions with the structure $-P$, and which indicate that a given state of affairs (P) is not the case. Thus, they can be paraphrased as follows.

(57) a. It is not the case that money goes with pennants. / Money does not go with pennants.

    b. It is not the case that privileges go with pennants. / Privileges do not go with pennants.

    c. It is not the case that someone has done something of benefit.

Neither this explanation nor a more functionally oriented analysis stating that we are facing negative statements that contain constituent-negation reveals why the reader perceives the negative utterances as being slightly humorous. To be able to capture this aspect, it is necessary to make reference to how discourse is based on information that is shared by interlocutors in a communicative situation. The negative utterances in (56) are slightly humorous because they go against the commonly held assumption that prizes, even if financially irrelevant, are very highly valued because they provide prestige and status. The negative statement obviously defeats a previously held assumption about the way certain things are valued in our society. To capture this interpretation, we need a framework to address the following questions: in general terms (1) what is the role of knowledge in text processing; and, more specifically, (2) what is the function of negation in discourse; to answer question (2) it is necessary to determine what the relationship is between the negative and the corresponding affirmative term. These questions are answered throughout the rest of this book.

Example (58) illustrates an example of contradiction in *Catch-22*.

(58) Group Headquarters was alarmed, for there was no telling what people might find out once they felt free to ask whatever questions they wanted to. Colonel Cathcart sent Colonel Korn to stop it, and Colonel Korn succeeded with a rule governing the asking of questions. Colonel Korn's rule was a stroke of genius, Colonel Korn explained in his report to Colonel Cathcart. Under Colonel Korn's rule, the only people permitted to ask questions were those who never did. Soon the only people attending were those who never asked questions, and the sessions were discontinued altogether, since Clevinger, the corporal and Colonel Korn agreed that it was neither possible nor necessary to educate people who never questioned anything. (p. 49)

Whereas the first negative proposition in the extract (1) *for there was no telling what people might find out once they felt free to ask* illustrates a typical use of negation, in the sense that it does not violate any laws of logic or rules of acceptability, and can be grouped with the negatives discussed in (56), the rest of the negative propositions in this extract are involved, directly or indirectly, in contradiction. Sentence (2) *the only people permitted to ask questions were those who never did* can be said to incur in contradiction, as we may understand it to be paraphrased as follows:

(59) You can ask questions if, and only if, you don't ask questions.

The proposition paraphrased in (59) does not have, strictly speaking, the form of a logical contradiction as it was described in Section 2.2.1.1.; however, it is clear that we are facing a complex structure that conveys contradictory information. This seems to point to the need to consider contradiction in discourse in broader terms than in logic, so as to account for occurrences such as those in extract (58).

Further, the contradictory meanings that emerge in the example discussed here may not be interpreted in the terms proposed by Sperber and Wilson (1986), that is, by favoring one of the terms. They also cannot be interpreted by assigning validity to each term in different domains, as proposed by Escandell (1990). Extract (60) contains deliberate irresolvable contradictions.

The following structures involving negation are a development of the rule invented by Colonel Korn: (3) *Soon the only people attending were those who never asked questions*. In this case, there is no overt contradiction, however, it implicitly reinforces the previous statement because it can be paraphrased as follows.

> (60) Only the people who don't ask questions go to the educational sessions where one asks questions.

This interpretation of the sentence is recoverable by means of our knowledge of the world and of how educational sessions are supposed to work. It is this kind of information, in particular, the assumptions about how educational sessions should be directed, that allows us to identify the oddity of the last two clauses involving negation: (4) *it was neither possible nor necessary to educate people* (5) *who never questioned anything*. These clauses are not logical contradictions either, and they are grammatically acceptable; however, there is something odd about them. Their oddity lies in the fact that they deny something that is implicitly recoverable from the previous discourse, and which, as readers, we assume on the basis of our knowledge of the world; namely, that if higher officers have started the educational sessions it is because they thought it was necessary for the soldiers. In this light, (58) is outrageous because it denies the possibility of asking questions precisely to those people who are interested in asking them.

This reasoning can lead to considering contradiction as a discourse phenomenon that cannot be restricted to its logical definition, as pointed out previously. In this sense, in extract (58), it becomes obvious that what is behind the Colonel's new rule is a contradiction within the Colonel himself, as he has first offered the educational sessions and now wishes to cancel them. The change in the officer's attitude reveals the kind of double moral the officers are playing with, where, on the one hand, they think educational sessions may be necessary, but, on the other, they do their utmost to prevent the soldiers from learning too much. This reasoning leads to the cliché that knowing is power and ignorance is submission, a pattern that is promoted by the officers.

Turning to the linguistic account of this phenomenon, the perception of the critical underpinning in this passage and others of a similar kind can be said to arise by means of conversational implicature. This process of interpretation of problematic propositions as meaningful is precisely described by Werth (1995c, p. 92) as a process where an apparently incoherent proposition is interpreted as conversational implicature and incremented into the common ground of the discourse as metaphorical, ironic, and so forth.

To sum up, extract (58) illustrates two important points: (1) contradiction is perceived as meaningful, in this case by interpreting that further meaning is conveyed by means of conversational implicature; and (2) contradiction in discourse needs to be considered in broader terms than the strict logical sense so as to encompass the contradiction of implicit meanings in discourse. We return to the discussion of these aspects in the sections devoted to contradiction in Chapter 5.

2.5.   DISCOURSE-PRAGMATIC APPROACHES TO NEGATION

In the following sections, the contributions made by different authors on negation within discourse-pragmatic frameworks are discussed.

2.5.1.   Negation in Discourse

The traditional view of negation as a logical operator that reverses the truth value of a proposition, has conditioned its status as a semantic concept in much of the literature in the field. Research in pragmatics and related disciplines, however, has incorporated negation as an element that determines the creation of specific discourse pragmatic functions, such as denial. A negative utterance can be part of any functional or pragmatic classification just in the same way as affirmative utterances are, as long as their different properties are accounted for. For instance, negative utterances are used as illustrations of specific speech act functions in Vanderveken's (1991) classification of speech acts.

Unfortunately, however, most work on negation on these terms has a semantic orientation, and the examples are typically isolated sentences. Very little work has been carried out on the discourse and pragmatic properties of negation in context. The pioneer work is Jespersen's (1917/1966) monograph; although this work does not deal with negation from a discourse perspective as it is currently understood, many of the author's intuitions on the use and properties of negation are pragmatically based. The most extensive works in the field are those by Givón (1978, 1979, 1984, 1989, 1993) and Tottie (1982, 1991). Whereas the former develops a framework of negation based on the notion of negation as a propositional modality and on the cognitive properties of negative states and events, the latter carries out a detailed computer-based study of the variants of negation in English speech and writing. Other authors have also contributed to the field, either by devoting sections of grammars, or of broader studies, to the functions of negation (Downing & Locke, 1992; Halliday, 1994; Horn, 1989; Lyons, 1977; Quirk, Greenbaum, & Svartvik,

1985, Werth, 1995c/1999), or by focusing on specific aspects of its use (Jordan, 1998; Leinfeller, 1994; Pagano, 1994; van der Sandt, 1991). Thus, Leinfeller (1994) considers the rhetorical properties of negation as a means of foregrounding and of establishing cohesion. The cohesive function of negation, in particular of lexical negation in discourse, is also mentioned by other authors (Halliday & Hasan, 1985; Lyons, 1977; Werth, 1984). The notion of negation as involving the denial of an assumption or a defeated expectation has also been the focus of attention of many studies, although the majority have been concerned with the psychological aspects of the processing of negative terms or sentences (Clark, 1976; Clark & Clark, 1977), rather than on the processing of negatives in discourse, be it spoken or written.

2.5.2. Illocutionary Acts Performed by Negative Utterances

Traditional theories of speech acts are not directly concerned with negation because negation is seen as the logical operator that is applied to an utterance in order to yield a complex speech act. Searle (1969, pp. 32–33) points out that negation can operate either on a proposition or on the force of a proposition. This distinction is illustrated in the following formulas and examples.

> (61) a. $F(P)$ I don't promise to come.
> b. $F(-P)$ I promise not to come.

As Lyons (1977, p. 769) points out, only negative utterances like (61)b constitute an illocutionary act of promising, whereas (61)a is rather a statement concerning the refusal to make a promise. As such, utterances like (61)a also perform illocutionary acts, defined by Lyons (1977, p. 770) as "acts of non-commitment." These acts need to be distinguished from not saying anything and from making descriptive statements. These kinds of acts are not considered by speech act theory, which concentrates on acts to which commitment is shown. Noncommittal acts are discoursively very significant, however, as Lyons observes (1977), because by refraining from committing oneself to the truth of a proposition P, one may be implying that P is actually true. Consider (62):

> (62) I *can't tell* whether he's crazy or not, I don't know him well enough.

By uttering (62), a speaker may be implying that the other person is crazy but he or she does not want to commit to asserting it.

A closely related issue is the status of reported speech acts, which can be problematic in a similar way. In the light of what has been said above, they can indicate noncommittal within the reported act. This is illustrated in (63)a and (63)b, from Downing and Locke (1992, p. 183) and Tottie (1991, p. 35) respectively.

(63) a. He didn't promise to come.
   b. He asked if she knew.

Example (63)a will be considered as the reported noncommittal to a promise, and (63)b as reported question. Functionally, both have the illocutionary force of assertives.

A further issue concerning the status of negatives as illocutionary acts is the fact that illocutionary acts can be conveyed both directly and indirectly. Examples in (64) show the difference between a direct act, in (64)a, or an indirect act, as in (64)b.

(64) a. I didn't say that.   (speech act of denial)
   b. Why don't you come along?   (indirect speech act: suggestion)

In Vanderveken's (1991) work, we find a classification of speech acts that accounts for negative utterances. His categories are exemplified by affirmative and negative utterances, where relevant. Vanderveken (1991) establishes a semantic-based classification of speech acts into five main groups: assertives, commissives, directives, expressives, and declaratives. Here, the category of *assertives* replaces Searle's (1969) *representatives*. Vanderveken also follows Searle (1975, p. 22) in considering that there are verbs that can perform more than one illocutionary force and others that are consistently ambiguous between two forms (Vanderveken, p. 168). Consequently, there are verbs that appear in more than one of the categories developed. The following is a list the speech act types that are illustrative of cases where negation is a component of the illocutionary act and are exemplified by the author by means of a negative sentence.

1. In the group of assertives, Vanderveken (1991, p. 169–181) includes the following negative acts: negate, deny, correct, disclaim, disagree, dissent, and object;
2.. Among the commissives, reject, refuse, and renounce;
3. Within directives, forbid, prohibit, and interdict;
4. Among declaratives, renounce, deny, and disapprove; and
5. Among expressives, disapprove.

This classification shows a predominance of assertive speech act types, including the categories that are most frequently mentioned by other authors, namely, negative statement, denial, and correction. It also reveals the ambiguity or multiple membership of certain verbs, such as deny and disapprove. Thus, deny is defined as follows (Vanderveken, 1991, p. 170).

"Deny" is systematically both assertive and declarative. In the assertive sense to deny a proposition is to negate that proposition by asserting the contrary or opposite proposition. There is generally, perhaps always, a preparatory condition to the effect that the denial is a denial of something that has been affirmed. Further, while virtually any claim may be negated, denial seems to be related to matters of some importance and perhaps also related to accusation (further preparatory conditions). I may *negate* a claim that it is snowing outside by saying that it is not snowing, but it would take special contexual factors for me to want to *deny* it. On the other hand, I would naturally deny a (false) assertion that I had neglected to inform you of a contractual deadline.

The distinctions established here between *negate* and *deny*, bring together some of the main issues discussed by other authors, and which concern the question of whether the two illocutionary acts are variants of the same category, or whether they constitute separate categories. Although there are authors who consider negative statement and denial as separate categories (Brown, 1973, p. 17; van der Sandt, 1991, p. 331; Vanderveken, 1991, p. 170), there seems to be a tendency to consider them as variants of the same pragmatic function (Clark & Clark, 1977, p. 98; Givón, 1993, p. 190; Lyons, 1977, p. 777; Pagano, 1994, pp. 250–252; Tottie, 1991, p. 22; Wason, 1965, p. 7). This phenomenon is most clearly illustrated in the widespread use of the term *denial* to stand for the main function of negative utterances in most of the work of the authors mentioned. The reasons for this tendency are obviously related to the use of negation in discourse, a view that is not captured by semantic-based approaches to speech acts, which tend to deal with illocutionary acts of utterances in isolation. Similarly, a speech act account of this type cannot deal with textual related functions of negation, like contrast and contradiction, which are also mentioned as relevant discourse-pragmatic functions of negation by other authors (see for example Jespersen, 1917/1966, pp. 4–5).

Continuing with the discussion on the classification provided by Vanderveken (1991), it has to be pointed out that, although the speech acts listed can be recognized as prototypical speech acts of negation in that the illocutionary act is typically associated with the utterance of a negative sentence, there are other speech acts that are not typically or necessarily negative, but to which the negative operator can be applied. This can take place both in direct and indirect speech acts. Thus, the most common type of these acts is probably that of questions, which Vanderveken (1991, p. 190) lists under directives, but there are others, such as warning, reminding, advice, caution, hypothesis, swear, and so forth, which can be manifested by means of a negative proposition. Some of these illocutionary functions are illustrated under (65) as follows.

(65) a. Didn't he give you his address?   (question)
      b. Don't go near the cliff!   (warning)
      c. I wouldn't ring her before Tuesday. (advice)

It is particularly important to observe that negative utterances are not necessarily expressions of challenging or disrupting illocutionary acts, contrary to Givón's (1993, p. 188) observation that "negation is a confrontational, challenging speech act." This is pointed out by Downing (1995, p. 233) and Downing and Locke (1992, pp. 182–184), and even Givón (1993, p. 195), who admits that negation is sometimes used as a polite down-toner. Downing and Locke (1992, p. 184), illustrate the function of negatives as polite hedges. Both politeness-related uses are illustrated in the examples under (66).

(66) a. I really can't say...   (polite hedge)
     b. Wouldn't it be better if...   (polite suggestion)

Similarly, it is also the case that although negative utterances typically perform certain illocutionary acts, such as contradicting and correcting, this does not mean that all of these illocutionary acts are performed exclusively by negative utterances. Thus, Clark (1976, p. 35) observes that the illocutionary functions of agreement and contradiction can be carried out both by affirmative and negative sentences, as shown in the examples reproduced in (67) and (68) (Clark, 1976).[27]

(67) a. So Mary has been here all day? Indeed, she has.
     b. So Mary hasn't been here all day? Indeed, she hasn't.

(68) a. So Mary has been here all day? I'm sorry, she hasn't.
     b. So Mary hasn't been here all day. I'm sorry, she has.

Example (67) illustrates an agreement expressed by an affirmative in (67)a and a negative in (67)b. Example (68)a illustrates a negative contradicting an expectation, however, in (68)b is an affirmative carrying out this function. The notion of contradiction used here is obviously function-based and different from the formal approach of standard logic, where contradiction is found in a single proposition.

Many authors agree that negative utterances are typically used to make negative statements. Thus, Quirk and colleagues (1985, p. 179) consider this to be the main function of a negative utterance, which, however, can also carry out other functions, such as asking tactful questions, uttering exclamations, and giving commands. Similarly, Downing and Locke (1992, pp. 183–184) mention the following illocutionary acts that can be carried out by negative clauses: negative statement, question (indicated by a negative statement with rising intonation), request, exclamation, directive, promise, and polite hedge.[28] Downing and Locke also point out (1992, p. 184) that transferred negation takes place with certain verbs expressing mental process. Horn (1989, p. 202) stresses the disparity between the logical symmetry and the functional asymmetry of affirmation and negation, and points out that the main functions of negative sentences are to correct and contradict. Horn does not, however, deal with the functional aspects of negatives in discourse, as most of Horn's work is devoted to issues regarding philosophical, psychological, and semanticopragmatic questions of negation with examples that are not contextualized.

2.5.3.   Discourse Functions of Negation: Tottie's Denials and Rejections

The problems of defining the functions of negative utterances arise with regard to the organization and labeling of the different categories, a question already introduced in Section 2.5.1. It seems to be widely accepted that there are at least three main functions that can be carried out by negative utterances, summed up by Brown (1973, p. 17) into nonexistence, denial, and rejection. As argued in Section 2.5.1. previously, there seems to be a tendency to classify nonexistents and denials together as forming a broader class of negative utterances that has a common discourse function: that of denying a proposition that was either explicitly stated or which expressed an assumption or an expectation that is being defeated.

Tottie (1991, p. 22) establishes a classification of negative utterances based on the distinction between *denials* and *rejections*. Among denials, Tottie distinguishes between explicit and implicit forms. This distinction, as Pagano (1994, p. 252) points out, rather misleadingly refers to whether or not the corresponding affirmative proposition has been explicitly stated in previous discourse, and not to the character of the negative utterance. Thus, although the notion of explicit denial corresponds to the prototypical sense applied to this category, implicit denial is used by Tottie to mean negative statement or negative assertion. The distinction is illustrated in examples (69) and (70).

    (69) a. I've lost my wallet...
           b. No, you haven't, I found it under the sofa this morning.

    (70) a. What do you think about Jack?
           b. Well, he's not the kind of person I'd go out with...

Example (69) exemplifies an explicit denial, which is easily recognized by the possibility of recovering ellipted constituents in the negative clause. Example (70) illustrates an implicit denial, where there is no explicit utterance in the discourse that is being denied but, rather, an implicit assumption that the speaker imagines the hearer might hold and that is being denied.

Tottie (1991, p. 21) argues that the two types of negative utterance can be considered to belong to the same category on pragmatic grounds, in the sense that they both deny the truth value of propositions, some of which are expressed, while others are inferred contextually. This position differs from the traditional distinction between these two categories. Tottie (1991) argues that the reasons for postulating two different classes of negative statement are based on semantic and psycholinguistic reasons. Tottie makes the following observation on this point.

> It seems clear that the production of sentences expressing the nonexistence or nonpresence of objects is characteristic of a particular stage in the acquisition of language by children. One might also claim that nonexistence forms a separate

semantic category expressing the absence of an object rather than denying the truth of a proposition. (Tottie, 1991, p. 21)

It appears that the classification defended by Tottie (1991) foregrounds the type of link that is established between the negative utterance and the corresponding affirmative proposition; however, it overlooks the communicative functions of the two different types of denial, which also provide a pragmatic perspective on these utterances. Thus, explicit denials are typical of interactive discourse, and the results in Tottie's (1991) study show that they do not appear in the written language. Implicit denials, however, which are the most widespread type in general, occur both in speech and writing, and their interpersonal component is less obvious than in the case of explicit denials.

With regard to Tottie's (1991) category of rejections, she argues that both rejections and refusals are variants of the same category. Tottie distinguishes this class from that of denials on several grounds: (1) rejections, unlike denials, are not found only in language because a rejection and a refusal can be expressed by means of gestures or body language; and (2) rejections contain a component of "volition" that is not present in denials. Although the first point may reflect a real distinction between the linguistic use of negation and the use of negative acts in general terms, the second point is arguable on the grounds mentioned by Pagano (1994, p. 251); it is not the presence or absence of volition, which is probably present in all human communication, but, rather, the predominance of the ideational or the interpersonal components in each of the acts. Thus, denials are predominantly ideational because their function is to deny the truth value of a proposition. Rejections, however, are predominantly interpersonal.

2.5.4. Negation in English Speech and Writing

Hardly any work has been carried out on the variations of negation and the functions of negative utterances in context, although some authors mention small scale studies of the frequencies of negative items or clauses within a corpus. Givón (1993, p. 191) provides the frequencies of negative and affirmative clauses in a small sample of English narrative texts that includes two text types, academic writing and fiction. The results show that the affirmative is by far the most frequent clause type (95% of occurrences in academic writing and 88% in fiction). Negative clauses have significantly lower frequencies, with a higher percentage in fiction (5% in academic discourse, 12% in fiction). Givón (1993) accounts for the higher percentage of negatives in fiction by explaining that fiction also contains conversation, which is an interactive mode, while academic writing does not. Tottie's (1991) study concentrates on the differences of occurrence of *not-negation* and *no-negation* in English speech and writing. The corpora used by this author do not include fiction, a genre which the author avoids because of the problems involved in analyzing a hybrid mode. A preliminary analysis (Tottie, 1991, p. 17) shows that negatives are more than twice as frequent in spoken than in written language (27.6% of occurrences in speech versus 12.8% of occurrences in writing, counted as number of negative items per 1,000 words). The author considers that the presence of the pragmatic signal *no* in conversation is not enough to account for this difference, which, she suggests, is produced by a combination of factors that are directly linked to the interactive character of spoken discourse. Thus, Tottie proposes a classification of discourse categories for negation in spoken language where most of the categories do not occur in written language. The discourse functions proposed by Tottie (1991, p. 37) are the following: explicit and implicit denials, rejections, questions, imperatives, supports, and repetitions. Of these, only implicit denials and, arguably, repetitions can occur in written language. By supports, Tottie (1991, p. 34) means "listeners' signals that information has been received, accepted, and agreed upon," as in example (71).

> (71) a. it wasn't typical.
>   b. no.

Repetitions are defined by Tottie (1991, p. 36) in the following terms: "Repetitions may also be used as a floor-holding device, to prevent another speaker from taking over the turn. This use is hard to distinguish from repetitions due to performance factors, when the speaker repeats old material while trying to continue by adding new words and phrases." The frequencies of the different types are the following: implicit denials are the most frequent category (63%), followed by explicit denials (14%), questions (8%), supports (8%), rejections (2%), and repetitions (4%). There was only one occurrence of an imperative and it was not statistically significant.

With regard to the categories themselves, they are undoubtedly adequate for the analysis of conversation. Certain observations, however, have to be made regarding the terminology used and the organization. Decisions regarding terminology should be consistent within a classification; but this is not always the case, thus, Tottie (1991, p. 37) uses terms naming illocutionary acts, such as denial and rejection, and other discourse functions, such as support and repetition, but uses the term *imperative*, which names the mood structure, instead of *order* or *command* or *directive*, which would be more adequate in the classification. Similarly, the classification as it stands does not establish a difference between what can be called *full illocutionary acts*, like denials, rejections, questions, and imperatives, and other functions like supports and repetitions, which have a different discourse function, as Tottie (1991) points out. Finally, although the term *support* is explained very clearly by the author (Tottie, 1991, p. 37), it is useful to point out that this notion of *support* is only indirectly related to the notion of *support* in other works in discourse analysis, such as Burton (1980), where acts in conversation are divided into supports and challenges.[29] This is important with regard to negation because, as discussed previously, not all negatives express challenging illocutionary acts, consequently, the fact that there is a class named *support* does not imply that the rest of the categories are challenges.

Tottie's (1991) study also deals with the variations of not-negation and no-negation in speech and writing. The results (1991, p. 46) show that not-negation is the most frequent alternative in both varieties (73% in spoken language and 67% in written language). These results also show that affixal negation is more frequent in writing (33% versus only 8% in conversation).[30]

2.5.5. Negation and the Denial of Background Information

Discourse-based approaches to negation tend to focus on the relation between the negative utterance and a corresponding affirmative, which has either been explicitly mentioned in previous discourse or expresses an assumption that is denied or an expectation that is being defeated (Clark & Clark, 1977, p. 98; Givón, p. 189; Horn, 1989; Jespersen, 1917/1966, p. 82; Jordan, 1998; Leinfeller, 1994; Pagano, 1994; Wason, 1965).[31] Thus, Wason (1965, p. 7) defines the function of negative statements as follows: "In assertive discourse the function of such statements is generally to emphasize that a fact is contrary to expectation. The *subjective* context for their utterance is the assumption that another person, or persons, might classify a fact wrongly." This applies to situations that are not necessarily interactive, such as those described in experiments by Wason (1965) and Clark (1976), where the production of negatives corresponds to the defeated expectation in the speaker regarding the presence of an object that is not present. It also applies to interactive situations, both in spoken and written interaction. In spoken interaction, the process is more obvious and it can be summarized by Givón's (1993, p. 190) exemplification of how background assumptions differ in affirmative and negative utterances.[32]

Affirm-assertion:	The hearer does not know.
	The speaker knows.
Neg-assertion:	The hearer knows wrong.
	The speaker knows better.

Similarly, other authors (Horn, 1989; Pagano, 1994, p. 254) also stress the crucial relation between the utterance of the negative sentence and the idea that the speaker, by virtue of the communicative principle, wants to correct or prevent a supposedly wrong assumption that the hearer might hold. This process is described by Pagano with regard to written language as follows.

the writer creates a picture of the reader, who thus becomes an "ideal reader," and attributes to this reader certain experience, knowledge, opinions and beliefs on the basis of which the writer builds his/her message. . . . As the writer somehow assumes what the reader's questions and expectations are, s/he tries to provide information about these. Therefore, in cases where certain information is non-existent, the writer can report that by means of denials of what was expected. (Pagano, 1994, p. 253).

The fact that negative utterances are used to deny previously held assumptions is illustrated in the examples under (72).

(72) a. There's no milk left!
    b. A: We could drive to the center this afternoon.
       B: I've got no petrol.
    c. Visitors are requested not to feed the animals.

Example (72)a illustrates the use of a negative utterance to indicate nonexistence functioning as the denial of an expectation that there should be milk in the refrigerator. This kind of utterance is not necessarily interactive, as its main function is mainly descriptive of a state of affairs. Example (72)b is an exchange where speaker B denies the assumption that she imagines A holds regarding her car; and (72)c is a written notice typically found in zoos, which denies the assumption a visitor might have that it is possible, or that one is allowed, to feed the animals.

Jordan (1998, pp. 712–713) rejects the idea that only negation has a presuppositional nature and argues that affirmative utterances can be used when the negative is presupposed. An example is the sign "For sale," which presupposes that houses usually are not for sale unless explicitly stated. A crucial point in the argumentation is that the presuppositions evoked both by affirmative and negative utterances, such as those in the "For sale" sign and other typical signs such as "No entry" or "No parking," are typically of a cultural nature, so that it does not make sense to talk about the denial of an implicit proposition. Although the latter point is true of certain utterances, there is still a difference between the kind of presuppositions or background knowledge evoked by the affirmative signs such as "For sale" and the negative signs such as "No entry." Although the negative defeats an expectation— the expectation that any door can be used to enter a place—the affirmative does not because one does not expect, as a general rule, that any house will be for sale at any time. Jordan (1998) further proposes a model for the analysis of the textual functions of negation based on the relation between the negative utterance and other preceding and following utterances. Thus, Jordan (1998) distinguishes between textual patterns consisting of one, two, or more part-structures, where negation is the articulating utterance. For example, three-part patterns can consist of an affirmative utterance followed by a denial which is then followed by a second affirmative utterance that explains the reason for the previous denial.

Pagano (1994, p. 258) proposes a classification of denials in written texts in four different subtypes, depending on the type of relationship established between the denial and the proposition that is being denied. The four types are the following:

1. Denials of background information;

2. Denials of text processed information;
3. Unfulfilled expectations; and
4. Contrasts.

While the first three categories are clear, and the examples provided by the author illustrate the definitions, the last one is problematic. The category of *contrasts* is really a different category altogether from the others in the classification. A relationship of contrast can be expressed by prosodic, semantic and/or structural means. Pagano (1994, p. 263) follows a semantic approach, which leads to a classification as a contrast the following example under (73).

(73) For past generations, lifestyle was the leading pharmacopeia. *They had no antibiotics, no cures for infectious disease.* They had to rely on their manner of living to preserve their health. (Pagano, 1994, p. 263)

Pagano (1994) writes about example (73): "Here, there is an implicit comparison between the past and the present, and the differences are pointed out." It is, however, not at all clear that this and other examples should be contrasts, whereas others that are classified under any of the three other categories should not, especially if they are introduced by overt markers of contrast, such as *but*. The first three categories approach negation from a cognitive perspective; whereas the notion of contrast is, rather, a textual-semantic category.[33] As such, contrast is usually indicated by structural-semantic and/or prosodic means; for instance, by the presence of explicitly contrastive words or structures or by the application of contrastive stress and focus (see Werth, 1984, Chapter 7). This means that any of the three cognitive categories, denial of background information, denial of text-processed information, and denial of expectation, can be contrastive.

With regard to the three cognitive categories, intuitively, they systematize the types of denial that can be said to operate during the reading process. Thus, the denial of background information denies assumptions about shared beliefs or other cultural knowledge; the denial of text-processed information denies information that has been previously introduced in discourse or as an anticipation of what is going to be said. This function is particularly important from the point of view of the contribution of negation to the cohesion and coherence of a text, as it establishes connections with previous and subsequent discourse. Finally, denials can also indicate unfulfilled expectations. The distinction between the categories is not clear cut, particularly the notion of denial of expectation. An expectation can be created by activation of certain shared knowledge or background information from elements present in the context of the situation, or it can be created by means of the activation of background knowledge in text-processed information. In this sense, it is not clear that it constitutes a separate category.

A further problematic issue regarding the use and processing of negatives concerns the criteria for pragmatic acceptability of negative utterances in a context.

There are two ways of establishing the appropriateness of negative utterances in a context: one as established by frame-semantic or schema-theoretical principles (see Fillmore, 1982, 1985; Pagano, 1994; Shanon, 1981),[34] and the other, as established by the ontological properties of negative states and events (Givón, 1993). The first approach is discussed in Chapter 4, and the second approach is discussed in Section 2.6 as follows.

2.6.   GIVÓN'S FUNCTIONAL-PRAGMATIC THEORY OF NEGATION

The following sections are devoted to a discussion of Givón's (1978, 1979, 1984, 1989, 1993) theory of negation. It is one of the few fully developed discourse-pragmatic theories on negation in English, and some of the assumptions are incorporated into the text world model proposed in Chapter 5 for the analysis of negation in the novel *Catch-22*. Givón is concerned with several crucial aspects of negation, such as the marked character of negatives, the discourse presuppositional nature of negation, the role played by negation as a propositional modality, and the ontological properties of negative states and events. Each of these aspects is discussed in the following sections.

2.6.1.   The Marked Character of Negation

Negation is standardly considered to be the marked form in the polarity system. Some of the reasons for its markedness have been mentioned in Section 2.2.3. previously, where the cognitive principles that condition the production and processing of negative structures and negative lexical items are described. Givón (1979, pp. 115–130) explains the marked character of negation by virtue of the following features: (1) distributional restrictions; (2) syntactic conservatism; and (3) psychological complexity. With regard to the first aspect, the author (1979, p. 119) argues that marked structures have distributional restrictions that condition the freedom of elements that can be embedded in them and the possibility of embedding themselves in other structures. Among the structures where such restrictions apply with regard to the use of negation, Givón (1979) provides the following examples.

(74)   a.   Where did you leave the keys?
       b.   ? Where didn't you leave the keys?

(75)   a.   He continued to work.
       b.   ? He continued not to work.

(76)   a.   There used to be a story that went like this...
       b.   ? There didn't used to be a story that went like this...

The distributional restrictions of negation are pragmatically motivated and are rooted in its ontological properties, which are discussed in Section 2.6.5., and its uninformativity as compared to the affirmative.

With regard to the syntactic conservatism of negative structures, Givón (1979, p. 121) argues that "negative clauses, which are more presuppositional and carry less new information in discourse, will turn out to be more conservative with respect to elaborative diachronic change." Thus, in several languages, changes are introduced first in the affirmative and then afterwards in the negative, though sometimes only in a partial way.

Finally, with regard to psychological complexity, an issue that has already been discussed, Givón (1979, p. 131) stresses that the longer processing time required for negative terms with respect to positive ones reveals conceptual complexity and not structural complexity in negation. Furthermore, the author points out that this complexity is related to pragmatic factors and not to the logic of negation (1979). Thus, in strictly logical terms, the assignment of a positive or a negative value to the members of an opposition (the pairs like *high–low, strong–weak*, etc., discussed previously) is arbitrary. In language, however, this is not arbitrary, but, rather, "it reflects deep pragmatic and ontological facts about the way the human organism perceives and construes the universe" (Givón, 1979, p. 131).

Horn (1989) accounts for the marked status of negation as an interaction between two opposing principles: a quantity-based principle that requires speakers to be as informative as possible (where affirmative sentences are prototypically more informative than negative sentences) and a relevance-based principle "directing the speaker to omit anything irrelevant to the concerns of his interlocutor which might increase processing effort" (Horn, 1989, p. 201). Horn argues that negative statements are typically, though not necessarily, less specific and less informative than positive statements, and that this establishes a pragmatic asymmetrical relation between the positive and the negative. The asymmetry does not lie in the relation between a negative and a positive proposition but between speaker denial and assertion, thus revealing a difference between the logical symmetry of affirmation and negation and its functional asymmetry. Horn (1989, p. 203) summarizes the characteristics of negation as follows.

> Negatives . . . are by nature no more false than affirmatives, but prototypically they are psychologically harder and more loaded, epistemologically less specific and hence less valuable, emotively more inhibiting (or at least less highly valued), and pragmatically more difficult to use appropriately within an arbitrary discourse context. Not every negation is a speaker denial, nor is every speaker denial a linguistic negation, but the prototypic use of negation is indeed as denial of a proposition previously asserted, or subscribed to, or held plausible by, or at least mentioned by, someone relevant in the discourse context.

To this, the author (1989, p. 203) adds that the strong asymmetricalist position is "literally false but psychologically true."

2.6.2. Negation as Propositional Modality

Negation is one of the propositional modalities in language (see also Halliday, 1994, p. 88; Werth, 1995c, p. 376). As such, it can be placed at the end of a scale where positive assertion (yes) is at one end, and negative assertion (no) is at the other end, with modalized options in between (maybe/maybe not) (see Halliday, 1994, p. 89). Givón (1984, p. 319) argues that the nature of negation is more complex, and that it is a hybrid mode that shares properties with presuppositions, realis (or factual) assertion and irrealis (or nonfactual) assertion. Thus, negation can be said to occupy different places on the scale depending on the criteria that are taken as a reference point. According to Givón (1984) there are at least three scales on which the relative properties of negation may be observed.

1. From the perspective of propositional semantics, which is concerned with truth conditions, negation occupies one of the extreme points of the scale, where presupposition occupies the opposite end. As such, negation expresses the reversal of the truth value of a proposition.

    This is justified by the fact that there is a cline going from the information expressed in presuppositions, which is taken for granted as true, to that expressed by means of realis-assertion, which is strongly asserted as true, to irrealis assertion, which is weakly asserted as true, to negative assertion, which is strongly asserted as false (see Figure 2.1).

2. Givón (1984, p. 322) argues that scale in Figure 2.1 is misleading from the point of view of subjective certainty or strength of belief, and must be replaced by a scale where both realis assertion and negative assertion occupy the same place (see Figure 2.2).

    On this scale, negation, as a form of assertion, indicates a "mid-level of certainty a speaker may assign to his assertion that an event/state did not take place" (Givón, 1984).

3. From the perspective of discourse pragmatics, that is, the use of negation in context, negation seems to share certain properties with presupposition. This is the perspective of negation as an illocutionary act that denies an utterance that has previously been said in discourse, an assumption or a defeated

PRESUPPOSITION → REALIS-ASSERTION → IRREALIS ASSERTION → NEG-ASSERTION

FIGURE 2.1.

PRESUPPOSITION → REALIS-ASSERTION
NEG-ASSERTION → IRREALIS ASSERTION

FIGURE 2.2.

expectation (see Section 2.5.5.). On this scale, negation occupies the same position as presupposition (see Figure 2.3).

The similarities of negation with the mode of irrealis and with presupposition are discussed in Sections 2.6.3. and 2.6.4.

2.6.3.   Negation and Irrealis

Givón (1989, pp. 162–193) suggests that there is a similarity in the semantics of negation and irrealis regarding referential opacity (i.e., the possibility of having a nonreferential interpretation of NP arguments within opaque propositions). Givón argues that argument NPs can only be interpreted referentially if they are under the scope of realis-assertion or presupposition. Both irrealis assertion and negation, however, make it possible to have a nonreferential interpretation of the NP arguments under their scope. This can be illustrated by the following examples, from Givón (1989, p. 163).

(77) John saw *a movie*.   (R-asserted)
(= there's a *particular* movie that John saw)
(78) It is good that John saw *a movie*.   (Presupposed)
(= there's a *particular* movie that John saw)

In examples (77) and (78), under realis-assertion and presupposition above, the NP *a movie* is interpreted referentially, that is, it refers to a particular movie. Compare it with examples (79) and (80).

(79) John may go to see a movie tomorrow.   (Irr-asserted)
(i) There's a particular movie that John plans to see.
(ii) John plans to see some movie, though neither he
nor I have in mind any particular movie.

PRESUPPOSITION
NEG-ASSERTION → REALIS-ASSERTION → IRREALIS ASSERTION

FIGURE 2.3.

(80) John didn't see *a movie*.   (Neg-asserted)
      (i) There exists no movie such as John saw it.
      (ii) * There exists a *particular* movie such as John didn't
            see it.

In (79), under the scope of irrealis-assertion, both a referential (in ii) and a nonreferential interpretation are possible (in i). Finally, in (80), only a nonreferential interpretation is possible (as in i). Thus, Neg-assertion does not admit a nonreferential interpretation of an indefinite NP under its scope. To indicate referentiality, one must use a definite NP, as in (81).

   (81) John didn't see the movie.
         (= there's a *particular* movie such as John didn't see it).

Givón (1989) argues that the reason Neg-assertion does not accept a referential interpretation of indefinite NPs under its scope is probably related to the presuppositional nature of negation. If a negative statement or denial in discourse operates on a previously expressed affirmative proposition, or an assumption or expectation which is in some way familiar to both speaker and hearer, then the argument in the negative proposition, being co-referential with the one in the affirmative proposition, must be definite. As Givón (1989) explains, "If a proposition is familiar to the hearer, the identity of the referring arguments within a proposition must also be familiar to the hearer; the argument must then be definite." Furthermore, Givón (1984, p. 332) argues that it is a pragmatic factor that has to do with the fact that negative statements are not used to introduce new information in the discourse, or to introduce new referential participants, but, rather, to deny propositions or utterances that are already part of the common ground in the discourse situation.

This argument has very significant consequences on the way negation is perceived to function in discourse. Thus, these arguments can be summarized as follows. Negation allows two important operations (Givón, 1984, p. 332): (1) to eliminate the referential-indefinite interpretation of NPs that is possible under realis and irrealis; and (2) to make possible a nonreferential interpretation of NPs under realis.

The latter is easy to account for, according to Givón (1984, p. 332), who considers that "negation creates an explicitly nonexistent world, a feature it shares with irrealis, which creates a potential but not yet existing one." This feature of negation as a modality that projects a world is also mentioned by other authors (Leinfeller, 1994, p. 95; Pagano, 1994, p. 256) and is particularly interesting when considered from a text world perspective (see Werth, 1995c, p. 376), where negation enters the set of possible alternatives to the reality projected by what is the actual world. This is discussed in Chapters 3 and 5.

2.6.4. The Presuppositional Nature of Negation

The functional asymmetrical relation between the negative and the affirmative in discourse is expressed in terms of what Givón (1979, p. 93; 1993, p. 188) defines as the "presuppositional status of negative speech acts" (Givón, 1979, p. 93). The notion of presupposition here is obviously not the traditional semantic concept of presupposition, but rather, a discourse-based notion referring to the information that is already present in the common ground (Werth, 1995c, p. 91) or the context (Givón, 1989, p. 135). Context, in Givón's work (1989, p. 135) includes the following types of information:

a. Shared situational context:
   (i) deictically obvious information
   (ii) speaker as direct participant
b. Shared generic context
   (iii) information universally shared
   (iv) agreed-upon conventions, rules or games
   (v) divine revelation
c. Shared discourse context
   (vi) information which was asserted earlier in the discourse by the speaker and the hearer did not then challenge it.

Discourse presuppositional information is used by the author to refer to information that is "known to, familiar to or otherwise unlikely to be challenged by the hearer" (1989, p.135).[35] Within this framework, Givón (1993, p. 189) specifies that "a negative assertion is made on the tacit assumption that the hearer has either heard about, believes in, is likely to take for granted, or is at least familiar with the corresponding affirmative proposition." Givón (1993, p. 188) provides the following examples.

(82) A: What's new?
      B: The President died.
      A: Oh, when? How?

(83) A: What's new?
      B: The President didn't die.
      A: Was he supposed to?

The contrast between (82) and (83) shows the different discourse functions of the affirmative and the negative structures. Whereas (82) illustrates an exchange where new information is introduced (*The President died*) and further information is required about the topic, (83) reveals that the introduction of new information by means of the negative form is made against the background of a corresponding affirmative proposition. In this case, it would be the assumption or belief shared by the speakers in (83) that the president was going to die, and now that expectation is being defeated.

Langacker (1991, p. 132) agrees that negation is conceptually dependent on the affirmative form because "it makes salient (through schematic) internal reference to the situation whose existence it denies." Langacker goes on to suggest that negation can be interpreted in terms of the notion of mental space used by Fauconnier (1985) in such a way that the function of negation may be that of specifying the absence of an entity from a given mental space that is introduced by means of the affirmative term. Langacker (1991, p. 134) describes the relationship between negation and affirmation in the following terms.

> With respect to a background conception in which some entity occupies a mental space, M, [negation] portrays as actual a situation in which that entity fails to appear in M. The missing entity is a process in the case of clausal negation, but that is not the only possibility; for example, when *no* is used to ground a nominal (as in *no cat* or *no luck*), the entity absent from M is a thing.

This view of negation operating on the background of a mental space is further developed in Chapter 5.

Givón (1993, p. 189) distinguishes among the following types of relationships that can be established between the negative structure and the corresponding affirmative.

1. The affirmative proposition can be explicitly stated in the previous discourse, either by the same speaker, or by a different speaker. This is illustrated in examples (84) and (85).

    (84) I asked John to lend me that book, but he didn't.

    (85) A: I thought you were coming along.
         B: No, I'm not. What made you think I would?

2. The affirmative proposition may not be expressed in previous discourse, (see, for example, Tottie's 1991 implicit denials). In this case, the negative can deny a background expectation or assumption in the hearer. This is illustrated in (86) (Givón 1993, p. 189):

    (86) A: So you didn't leave after all.
         (i) No, it turned out to be unnecessary.
         (ii) Who said I was going to leave?
         (iii) How did you know I was going to?

In (86), there are three different possible reactions to A's utterance. Each of them shows a different relationship between A's beliefs and B's beliefs or assumptions about the fact that B should be leaving or not. (i) Shows that both speakers share the same background assumptions, and they are being denied. (ii) Reveals that A has probably been misled and his wrong assumption is being corrected. (iii) Shows B's surprise at A's knowing the information, which is presented as true.

3. Background assumptions can also be part of the culturally shared information shared by speakers. Givón (1993, p. 189) compares the appropriateness of sentences like those in (87) and (88).

(87) a.   There was once a man who didn't have a head.
    b.   ? There was once a man who had a head.

(88) a.   ? There was once a man who didn't look like a frog.
    b.   There was once a man who looked like a frog.

Examples (87)a and (88)b are felicitous because they single out exceptions from the general norm (a man without a head versus the general norm of having a head, and a man who looks like a frog versus the general norm of humans not looking like frogs). Examples (87)b and (88)a, however, are inappropriate because they are tautological; they repeat the general norm and do not add any new information.

This leads us to the ontological properties of negative states and events, which are discussed in Section 2.6.5.

2.6.5.   The Ontology of Negative Events

Givón (1993, p. 190) accounts for the differences in appropriateness of use of affirmative and negative statements in discourse by establishing a similarity between this opposition and that of the *figure/ground* distinction in cognitive psychology.[36] Changes, or events, are less frequent and cognitively more salient than stasis, or nonevents. Thus, an event that is prototypically expressed by means of the affirmative, stands out against the background of stasis, or nonactivity; it expresses the counternorm against a background of normality. Being less frequent and more salient cognitively, events are more informative than nonevents.

Negation, as a linguistic phenomenon, can be viewed as "a play upon the norm" (Givón, 1993, p. 190). "It is used when—more rarely in communication—one establishes the event rather than inertia as ground. On such a background, the non-event becomes—temporarily, locally—more salient, the more informative" (Givón, 1993, p. 190). Givón (1993, p. 191) illustrates this point with the following examples.

(89) a. A man came into my office yesterday and said...
b. A man didn't come into my office yesterday and said...
c. Nobody came into my office yesterday and said...

As the author (Givón, 1993, p. 191) points out "The non-event . . . is pragmatically—and indeed grammatically—the oddest. This must be so because if an event did not occur *at all*, why should one bother to talk about a specific individual who 'participated' in that non-event?" Visits to one's office are rarer and less frequent than the times nobody visits one's office, so that visits, as events, are more salient, and thus more informative, than nonvisits, or nonevents. This explains why (89)b and (89)c would not be frequently used as discourse initials, unless there was a clear expectation that the contrary should have been the case. This means that utterances such as (89)b and (89)c will typically function as denials, contradictions, or corrections of previous utterances, as illustrated in the examples under (90).

(90) a. A: A man went yesterday into your office...
b. B1: A man didn't come yesterday into my office...
c. B2: Nobody came into my office yesterday...

2.7. AN APPLICATION OF GIVÓN'S FRAMEWORK TO NEGATION IN *Catch-22*

In the two extracts that follow, the notions of the presuppositional nature of negative sentences and of their ontological properties is discussed, along with a comment on their significance in the analysis of the function of negation in discourse.

2.7.1. The Presuppositional Nature of Negation

Example (91) is a good illustration of the presuppositional discourse properties of negative sentences.

(91) He gasped in utter amazement at the fantastic sight of the twelve flights of planes organized calmly into exact formation. The scene was too unexpected to be true. There were no planes spurting ahead with wounded, none lagging behind with damage. No distress flares smoked in the sky. No ship was missing but his own. For an instant he was paralyzed with a sensation of madness. Then he understood and almost wept at the irony. The explanation was simple: clouds had covered the target before the planes could bomb it, and the mission to Bologna was still to be flown. He was wrong. There had been no clouds. Bologna had been bombed. Bologna was a milk run. There had been no flak there at all. (p. 186)

The episode describes the reaction of Yossarian, the protagonist, when the planes that were supposed to bomb Bologna on a very dangerous mission, return undamaged. He is on the ground because he has managed to find an excuse not to fly the mission. We can divide the extract above into two sections, and consider the function of the negative statements within them. In the first part, up to *Bologna was still to be flown*, we have the following negative statements: (1) *There were no planes spurting ahead with wounded;* (2) *none lagging behind with damage*; (3) *No distress flares smoked in the sky*; and (4) *No ship was missing but his own.* These statements clearly have the function of denying an expectation held by Yossarian about the return of the planes that had gone on the mission to Bologna. None of this information has been expressed previously in the discourse, however, it is present implicitly as an assumption that is based on knowledge of the world, more specifically, on knowledge about events during a war: if planes go on a dangerous mission, it is to be expected that there will be damaged planes and wounded soldiers, but this expectation is not fulfilled. This phenomenon illustrates the notion of "discourse presuppositionality" discussed previously. The negative propositions presuppose that an expectation has been previously created that is now denied. The facts described make Yossarian infer that the weather conditions have prevented the mission from taking place, and lead him to develop a new assumption, which adapts to the changed situation. He believes that the mission still has to be run. The second part of the extract denies this second assumption, by means of stating explicitly that his assumption was wrong in the affirmative statement (5) *He was wrong*; and by explaining why he was wrong, in the negative statements (6) *There had been no clouds;* (7) *There had been no flak there at all.* Sentence (6) denies information that is present in the previous discourse (*clouds had covered the target*), and (7) denies the expectation that the enemy would defend themselves from the attack and counter-attack.

To sum up, the argument that negation has a presuppositional nature provides the means of accounting for one of the reasons why negative sentences are used in discourse. The example discussed here illustrates the point that negatives deny expectations and assumptions held by speakers that can be explicitly expressed in previous discourse or implicitly present in the common ground. It is also interesting to observe that there is a progression in the negative sentences of the extract, corresponding to the two different parts. This shows that the function of negation in discourse needs to be considered in the context of the preceding and following utterances because the isolation of the sentences would limit the possibilities of interpretation enormously.

2.7.2. The Ontology of Negative States and Events

In example (92), the significance of Givón's observations about the presuppositional nature of negation and of the ontology of negative properties and events is considered.

(92) "What the hell are you getting so upset about?" he asked her bewilderedly in a
tone of contrite amusement. "I thought you didn't believe in God." "I don't,"
she sobbed, bursting violently into tears. "But the God I don't believe in is a
good God, a just God, a merciful God. He's not the mean and stupid God you
make Him out to be." Yossarian laughed and turned her arms loose. "Let's have
a little more religious freedom between us," he proposed obligingly. "You don't
believe in the God you want to, and I won't believe in the God I want to. Is that
a deal?" That was the most illogical Thanksgiving he could ever remember
spending. (p. 231)

This extract is taken from a long conversation between Yossarian and Lieutenant
Scheisskopf's wife, his lover. The conversation has turned to religious matters by
means of angry comments and insults regarding religious systems in general, and
God in particular, on Yossarian's part. Lieutenant Scheisskopf's wife is affected by
his blasphemous utterances. The first negative utterance (1) *I thought you didn't
believe in God*, denies an assumption that Yossarian has held about his lover's
beliefs, on the grounds that she has previously told him she does not believe in God.
His comment is an utterance of surprise against her unexpected reaction. Surprisingly, lieutenant Scheisskopf's wife contradicts herself and confirms the original
assumption held by Yossarian, by means of (2) *I don't*. This negative utterance
simultaneously confirms and denies information conveyed previously in discourse.
The contradictory nature of this utterance is expanded by the following utterances,
which are unacceptable from a logical point of view: (3) *The God I don't believe
in is a just God, a merciful God* . . . . This utterance is an example of the polemical
structure where the presuppositions of a sentence are negated by the assertion (see
the discussion on this topic in Section 2.2.1.). The presupposed information is coded
in the NP Subject, where an entity, *God*, is defined in terms of a negative attribute,
I don't believe in. The word *believe* is crucial here because believing in God means
accepting the existence of an entity such as God. The predicate, however, *is a just
God, a merciful God*, which assigns positive properties to the entity God, states that
that entity exists. The result is a contradiction between the presupposed information
and the assertive content of the proposition. This view is confirmed by the utterance
(4) *He's not the mean and stupid God you make Him out to be*, where, again, the
speaker corrects Yossarian's assumption about God. The contradiction seems to
indicate that Lieutenant Scheisskopf's wife both believes and does not believe in
God at the same time. In Section 2.2.2.1., it was suggested that some contradictory
structures could be interpreted as meaningful by identifying two different domains
in which each of the contradictory terms is applicable. In this case, it would be two
internal mental domains of Lieutenant Scheisskopf's wife, which contain different
sets of beliefs, expectations, and wishes. With regard to the question of the
ontological features of structures of this kind, it can be concluded that these
contradictions are cognitively unstable structures, or, as Apter argues for cognitive
synergies, bi-stable structures; this means that a property is described in terms of

contradictory attributes but is perceived as bi-stable because each of the attributes tends to be focused on at different points in time. The incongruity and, possibly, the humorous character of the passage under discussion lies in the simultaneous presence of the contradictory attributes in one single structure.

To end this section, some comments about the foregrounding of negative events along the lines of Givón's observations are added. This is done by means of considering its applicability to the use of negation in the following example.

(93) Sharing a tent with a man who was crazy wasn't easy, but Nately didn't care. He was crazy, too, and had gone every free day to work on the officers' club that Yossarian had not helped build. Actually, there were many officers' clubs that Yossarian had not helped build, but he was proudest of the one on Pianosa. It was a sturdy and complex monument to his powers of determination. Yossarian never went there to help until it was finished—then he went there often, so pleased was he with the large, fine, rambling shingled building. It was truly a splendid structure, and Yossarian throbbed with a mighty sense of accomplishment each time he gazed at it and reflected that none of the work that had gone into it was his. (p. 28)

In (93), we have examples of negative clauses that foreground negative events. The first two negative clauses (1) *Sharing a tent with a man who was crazy wasn't easy,* and *but Nately didn't care,* carry out the standard function of foregrounding negative states for the purposes of correcting wrongly held assumptions in the reader (that it might be easy to live with a crazy man and that Nately should mind this fact). They connect back to previously processed information. The following negative clauses, however, are odd: (2) *the officers' club that Yossarian had not helped build*; (3) *Actually, there were many officers' clubs that Yossarian had not helped build*; (4) *Yossarian never went there to help until it was finished,* and (5) *and Yossarian throbbed with a mighty sense of accomplishment each time he gazed at it and reflected that none of the work that had gone into it was his.* These clauses also foreground negative states or events (*not build, not go, none of the work was his*) and, strictly speaking, they are not informative: why devote a whole paragraph to describing the actions not carried out by a character? This use of negation inverts the usual process in communication, where we are usually more interested about the actions of characters and how they contribute to the plot than in the nonactions, as these tend to form the background of the story. This paragraph shows the contrary phenomenon where the relevant event is the nonevent, the fact that a character has done absolutely nothing to contribute to the building of an officers' club, and that he is proud of it. In Chapter 5, further examples of this kind are discussed to show that the foregrounding of negative states and events as in extract (93) is a recurrent feature of the novel *Catch-22,* and, as such, it is significant for the understanding of the discourse of the novel.

Neither Givón's approach, however, nor any of the other frameworks discussed in the present chapter, account for the way in which the apparent uninformativity of extract (93) is recovered as ultimately informative by the reader, or how the contradictions in (92) are interpreted as meaningful. To account for these facts, we need a theory that will be able to tackle the question of how knowledge packages are activated and processed while reading. This theory may be schema theory, which is discussed in Chapter 4. Furthermore, the approaches discussed in this chapter are restricted to the analysis of isolated sentences or brief exchanges. This necessarily limits the explanatory capacity of even a powerful pragmatic theory, such as Givón's. Text world theories, discussed in Chapter 3, provide the means for dealing with negation throughout long discourse stretches. This sets the necessary grounds for a more detailed description of the properties of negation as a discourse phenomenon and for an explicit account of the relation between negative and affirmative within a text world framework.

2.8. CONCLUSION

In this chapter works on negation that are relevant to the discussion of negation in the novel *Catch-22* in Chapter 5 were reviewed. The definition of negation as a truth functional operator is inadequate for an analysis of negation in discourse. Different approaches to negation from grammatical and functional-pragmatic perspectives have been discussed so as to set the grounds for a proposal for the analysis of negation in a text world model, which is discussed in Chapters 3, 4, and 5. To sum up the observations made, I have tried to show (1) how negation can be said to have a presuppositional nature that is manifested in the links that are established discursively with previous assumptions and expectations; and (2) how negation foregrounds negative states and events in such a way that the nonbeing is focused on for reasons that are recovered contextually. These facts show that functional and ontological aspects of negation are primary in its interpretation as a discourse element. Subsequent chapters contribute to this line of thought by expanding on issues not developed by the frameworks discussed in this chapter, namely, the dynamic discourse function of negation and the way negation is processed in relation to stored knowledge.

NOTES

1. See also Bosque (1980) and Bustos (1986) for implicature-based accounts of negation in Spanish. Also see Oh & Dinneen (1979) for a collection of articles on classical themes related to the notion of presupposition.

2. For a detailed discussion of the supposed ambiguity of negation see Kempson (1975).

3. See McCawley (1981, pp. 62–64) for a discussion of how the truth table for negation is interpreted in logical terms. The main idea underlying this account is that, in logic, "a proposition and its negation must have opposite truth values" (p. 63). This means that if there is a proposition P, which describes a state of affairs and is assigned the truth value T, there must be another negative proposition not-P, which is assigned the truth value F for that state of affairs.

4. See McCawley (1981, pp. 67–69) for a discussion of a non truth-functional approach to negation. The truth functional or non-truth functional character of negation determines the nature of the connectors & and $if then$. McCawley points out that a non-truth functional approach to negation is ruled out by classical logic but it is worth consideration, as it can be derived from the application of the rules of inference in logic and can allow for the acceptance of cases where a proposition and its negation may be both true or both false. This may be the case of propositions that contain a false semantic presupposition, as in *The King of France is bald* and *The King of France is not bald* discussed in the present chapter. According to McCawley, "Under the narrow conception of falsehood, if a proposition and its negation both fail to be true, then both lack any truth value; under the broad conception of falsehood, if a proposition and its negation both fail to be true, then both are false." (p. 259)

5. For a discussion of this example within the model of text worlds, see the section devoted to negative accommodation in Chapter 5.

6. Compare Atlas and Levinson (1981, p. 32). These authors consider negation in natural language to be unambiguously of an external, widescope, sentential type, whereas "the usually preferred interpretation as a choice/narrowscope/predicate/internal negation is pragmatically induced."

7. There are other systems of logic that incorporate aspects such as phonological markers, however, I am here referring to traditional logic only.

8. Kempson (1975, pp. 95–100) points out the inadequacies of propositional logic in the distinction between descriptive negation (the term she uses for negative statement) and denial with reference to the problem of the ambiguity of negation. In the general literature on the discourse-pragmatic functions of negation, both "negative assertion" and "negative statement" are terms used indifferently to define a speech act of a descriptive type. It usually indicates the non-presence of something that was probably expected to be present. Denial, also a representative speech act, rejects the truth value of a proposition that either has been mentioned previously in discourse or has been implied. Compare Tottie (1991), for whom all negative representatives are denials, the difference being whether they deny a proposition explicitly mentioned or an implicit proposition.

9. Also see the section on contradiction in the analysis of the data in Chapter 5.

10. The inclusive meaning may be considered to be basic, and the exclusive meaning may be recovered by conversational implicature.

11. See the discussion in Chapter 5 on how the *either-or* disjunction reveals this kind of problem in the chaplain in *Catch-22*.

12. See section 2.3.4.2. on the distinction between contraries and contradictories.

13. The perspective on phenomena such as contradiction proposed by Eastern philosophy is significant if we compare it to the process of understanding a literary work where contradiction and anomalous logic are recursive, such as *Catch-22*. The process described by several authors (see Cook, 1994; Leech & Short, 1981; Norrick, 1986) whereby incon-

sistencies or incongruities at the literal level can be interpreted as meaningful at a higher level of interpretation, can be compared to the process followed in Eastern philosophy where the rejection of the literal value of specific propositions leads to an insight and understanding at a higher level.

14.   See also (Olson, 1997) for a study on the difficulties in conceptualizing absence in children. The author argues that "while negation is part of oral language, conceptualizing absence may be related to the invention of notations for negation" (p. 235).

15.   The symmetricalist view of negation argues that the relation between affirmative and negative is understood biunivocally. That is, the negative works upon the previous existence of the affirmative, but the affirmative also requires the existence of the negative in order to exist (see, for example, Jordan, 1998). This view is reformulated as an aesthetic theory of negativity that has also influenced current literary theories. Iser (1989), for example, claims that the notion of *negativity* in literature should be understood as the "unsaid," as what is absent in a text, which is the complementary of the written text itself.

16.   See Horn (1989, pp. 45–78) for a discussion of the views argued by symmetricalists and asymmetricalists, with regard to the question whether negation makes presupposes a corresponding affirmative form. See Voterra & Antinucci (1979) for an experiment on negation in child language which shows that the children produce denials that presuppose a corresponding affirmative statement.

17.   Compare Escandell's account of contradictory statements discussed in section 2.2.2.1.

18.   Jespersen (1917/1966) applied tests sporadically to identify some words, such as broad negatives, as negative. The application of tests was not carried out consistently and systematically, however, as is pointed out by McCawley (1995, p. 32) in his review of Jespersen (1917/1966).

19.   See Jespersen (1924/1961b, pp. 464–489) for a description of the most frequent negative prefixes in English.

20.   Clark (1976, p. 37) classifies negatives into four different types, according to two main criteria: whether or not they are full negatives (i.e., whether they are explicitly negative) and whether or not they are quantifier negatives.

21.   See Section 2.5.3. on variations in the use of negation in English speech and writing.

22.   For a detailed discussion of the constraints on the use of *no-negation* and *not-negation* see Tottie (1991).

23.   Huddleston (1984, p. 432) argues that in order for an utterance such as *John opened the door* to be true, the following entailments and presupposition (v) must be true:

(i)   Someone opened something.
(ii)   Someone opened the door.
(iii)   John opened something.
(iv)   John opened the door.
(v)   At the time prior to the time at issue, the door was closed.
etc.

The proposition *John opened the door* will be false when the set of conditions outlined above are not satisfied. From a strictly semantic point of view, however, negation does not specify which of the conditions is not satisfied in every possible situation. Huddleston (1984) argues that this information is provided pragmatically (by means of implications) or prosodically (by means of contrastive stress).

24. There are also adjuncts which tend to fall outside the scope of negation, as is observed by Huddleston (1984, p. 429). Thus, a sentence such as *He didn't do it because he was angry* is ambiguous, and the interpretation depends on whether the adjunct is understood to fall under the scope of negation or not.

25. This view seems to be grounded on the well-known Pollyanna principle, which states that the natural tendency is to focus on positive aspects, and not on negative ones.

26. For further discussion of the different types of relationships between opposites, see Lyons (1977, pp. 270–290), Werth (1984, pp. 151–165), Cruse (1986, Chapters 10, 11, and 12), and Horn (1989, pp. 35–45).

27. See also Lyons (1977, p. 771) and van der Sandt (1991, p. 331) for similar arguments concerning the fact that both affirmative and negative sentences can function as denials.

28. See also Tottie (1991, p. 37) for a classification of discourse functions of negation. I discuss this approach in section 2.5.4.

29. In Burton (1980, pp. 156–159) what Tottie (1991, p. 37) calls *supports* are classified as *accept* and *acknowledge* acts.

30. See the appendix with the frequencies of negation types in the novel *Catch-22*.

31. For a detailed discussion of the relation between negative utterances and the corresponding affirmative forms see section 2.6.4.

32. Givón's (1978, 1979, 1984, 1989, 1993) approach to negation is discussed in the last sections of the present chapter.

33. See Werth (1984, Chapter 7) for a discussion of contrast as one of the three functions of *emphasis* that contribute to the creation of coherence in texts.

34. The schema-theoretic account of the appropriateness of negatives in discourse has a cognitive basis that is obviously linked to the findings in cognitive psychology discussed in section 2.2.3.

35. Other authors (Clark, 1976; Clark & Clark, 1977), avoid using the term "presupposition" and use the term "supposition" to describe the phenomenon discussed by Givón.

36. For a discussion on the ontological properties of negative states see Section 2.2.3.1. and 2.3.4.1., on the cognitive properties expressed by means of lexical negation.

3

Negation and Text World Theory

Now that Yossarian looked back, it seemed that Nurse Cramer rather than the talkative Texan, had murdered the soldier in white; if she had not read the thermometer and reported what she had found, the soldier in white might still be lying there alive exactly as he had been lying there all along...
(Heller, 1961, p. 214)

3.1. INTRODUCTION

This chapter continues the discussion of negation within the frame of discourse studies by considering its function within a text world model. The frameworks discussed in this chapter, together with the approaches reviewed in Chapter 4, are intended to provide the basis for a dynamic discourse perspective on negation, a view that is not developed by any of the frameworks introduced in previous chapters. The first part of the chapter deals with Werth's (1995c/1999) text world model. The second part focuses on the notion of conflict in fictional worlds in Ryan (1991b). Werth's theory systematizes the observations regarding the cognitive and ontological properties of negation within a dynamic discourse model. Ryan's model provides the literary background for an analysis of the role of negation in the expression of conflict in the fictional world of *Catch-22*.

3.2.   POSSIBLE WORLDS, TEXT WORLDS,
FICTIONAL WORLDS

In recent decades, there has been a great deal of cross-fertilization between different disciplines, such as philosophy, logic, linguistics, cognitive psychology, and literary theory.[1] Text world theory has developed as a result of this integrative trend (see Allén, 1989; Bradley & Swartz, 1979; Doležel, 1976, 1989; Pavel, 1985, 1986; Ryan, 1985, 1991a, 1991b; Semino, 1995, 1997; Werth, 1994, 1995c/1999). The first step is to define what is understood by text world theory and the particular aspects of this model that are applied to the analysis of negation. Teleman (1989, p. 199) points out that the term *possible world*, from which the notion of text world derives, is often used in a vague and ambiguous way. It is associated mainly to two main trends in linguistic theory: propositional semantics and text theory. The notion of possible worlds, which was originally conceived by the philosopher Leibniz, was first introduced into modal semantics to account for the notions of necessity and possibility (see Lyons, 1995, pp. 118–119, van Dijk, 1977, pp. 29–30). Thus, in terms of propositional logic, we can say that a proposition P is necessarily true if P is true in all possible worlds, or, to put it in different terms, if it is true in "any situation we can imagine" (van Dijk, 1977, p. 29). Similarly, P is possible if there is at least one possible world (or situation) in which P is true. In these terms, "a possible world may be identified with the set of propositions that truly describe it" (Lyons, 1995, pp. 118–119). As van Dijk (1977) indicates, we can think of a possible world in more intuitive terms to represent a "situation" or a "state of affairs." As such, a possible world is an abstract construct about which a set of propositions are said to be true. Conversely, a proposition can also be defined in terms of its relation to the possible worlds where it is true; from this perspective, a proposition is the set of possible worlds at which it is satisfied.

Possible world logic has been particularly useful in the explanation of irrealis phenomena, such as the syntactic and semantic properties of modal verbs, counter-factual conditional clauses, and complement clauses of world-creating predicates (e.g., *want, believe*), as well as the tense and aspect systems (see Teleman, 1989, p. 199). Although, propositional logic has been extremely influential in the development of formal semantic theory and the systematization of truth conditional semantics, the application of possible world principles to text theory has focused on the idea of text as a mental construct (see Section 1.5.3. in Chapter 1). Thus, de Beaugrande (1980, p. 24) defines a text world as "the cognitive correlate of the knowledge conveyed and activated by a text in use. As such, it is in fact only present in the minds of language users." Similarly, de Beaugrande and Dressler (1981, p. 94) focus on the cognitive dimension of text processing as a crucial factor in the construction of text coherence.

> the combination of concepts and relations activated by a text can be envisioned as PROBLEM-SOLVING....Given some fuzzy unstable units of sense and content, text

users must build up a configuration of pathways among them to create a TEXTUAL WORLD. (de Beaugrande & Dressler, 1981, p. 94)

A text-based possible world perspective has led to the substitution of logical principles (e.g., logical necessity and logical possibility) by cognitive and epistemic principles more consonant with the process of text production and understanding.[2] This affects two main aspects of text.

1. Text processing, an aspect overlooked by formal possible world theories, which in text world theories is systematized according to principles such as coherence (see de Beaugrande, 1980; van Dijk & Kintsch, 1983) or coherence and co-operation (Werth, 1995c/1999), in connection with the way in which knowledge is stored, processed, and activated (de Beaugrande, 1980; Cook, 1994; van Dijk & Kintsch, 1983; Werth, 1995c/1999). This aspect is discussed in particular with relation to the processing of negation in the discussion that follows in this chapter and in Chapters 4 and 5.
2. The internal structure of text worlds, where the notion of *accessibility* replaces that of *possibility*, to explain differences in the internal structure of worlds or universes and how they diverge from a given actual world or central text world (see Sections under 3.4.).[3]

These principles have been applied to the analysis of fictional worlds by concentrating on two aspects (see Semino, 1995, p. 80): (1) the ontological status of fictional entities and of propositions in fictional discourse; and (2) the classification and description of fictional worlds. The latter aspect is discussed in the second part of this chapter, which deals with conflict within fictional worlds.

3.3.   WERTH'S TEXT WORLD THEORY

Werth's (1993, 1994, 1995a, 1995b, 1995c/1999) model is a proposal for a cognitive-based discourse framework that adopts notions from formal semantics, including possible world theory, which are adapted to a broader discourse perspective. Its characteristics as a discourse theory in general terms are established by means of identifying the *discourse* and the *text* as the units of analysis. This is because sentence-based approaches are considered by the author to provide only a limited range of insights into the properties of language and how it is used (Werth, 1995c, p. 6).

Thus, the author describes the proposal in the following terms.

all of semantics and pragmatics operates within a set of stacked cognitive spaces, termed "mental worlds." ... My argument for this, in a nutshell, is that uses of language presuppose occurrence in a context of situation, and that on top of this they also presuppose the existence of a conceptual domain of understanding, jointly constructed by the producer and the receiver(s). (Werth, 1995c, p. 26)

The quotation reveals the cognitive basis of communication in this framework, where interaction is considered to involve the activation and use of packages of knowledge and the construction of abstract conceptual structures of different levels of complexity. The general term *domain* is used henceforth to refer to these cognitive spaces or mental worlds that linguistically are defined by deictic and modal parameters, as described as follows.

3.3.1. Discourse World and Text World

As in other discourse theories, interaction is governed by a series of principles, among which those of informativeness, cooperativeness, and coherence are primary, as they regulate the negotiation of the discourse situation (Werth, 1995a, pp. 78–79). The definitions of *discourse* and *text* draw extensively from cognitive linguistics theories, in their stress on the cognitive aspects of the processing and the role of knowledge in communication and in the conception of texts as projected states of affairs or cognitive domains. Discourse and text are defined as follows.

> discourse: a deliberate and joint effort on the part of producer and recipient to build up a 'world' within which the propositions advanced are coherent and make complete sense....
> text world: a text world is a deictic space, defined initially by the discourse itself, and specifically by the deictic and referential elements in it. (Werth, 1995c, p. 95)

Further, discourse is defined as a *language event* and is identified with the *immediate situation*. The text is the language itself (Werth, 1995c, p. 26). Both discourse and text are constructs based on human experience (perception, memory, and imagination), which together make up a representation of discourse, and of reality itself, which is not directly accessible to human consciousness. Communication is described in cognitive terms, as a process where "speakers build up a repertoire of scenes which encapsulate their expectations about how particular situation-types will turn out." (Werth, 1995c, p. 172). The repetition of similar situation-types (manifested through particular text worlds) leads to the creation of *frames*, which are "conceptualisations of real-world phenomena" (Werth, 1995c, p. 181).[4] Thus, a text world will contain knowledge that is evoked by frames and knowledge that is contributed by the discourse situation. In particular, the discourse situation will provide specific variants to schematic frames (names, places, time, etc.), thus enriching the text world (Werth, 1995c, p. 259). Furthermore, the text world is not defined as a fixed entity independent from the interlocutors, rather, it is a dynamic phenomenon that undergoes constant change. In this sense, the text world is defined as "a representation of the cognitive space which the author and the reader are co-operating to form between them" (Werth, 1995b, p. 191). The space delimited by the text world is subject to change because both readers and the text itself change through time. Thus, the development of the text world can be compared to the succession of frames in a movie film, an analogy that foregrounds the view of text as ongoing process.

The way knowledge is organized plays a crucial role in the process of communication. The knowledge that is organized and negotiated in the discourse, which Werth (1995b, p. 91) defines as the common ground, is constituted by the set of expressed propositions in the discourse, plus the set of entailed and pragmatically connected propositions, which can be *potentially relevant* (p. 91), and some of which can be activated. The propositions refer to *possible situations* in the discourse world (p. 91), so that the propositions are consistent and cohere with the propositions that define the text world. This view constitutes a reformulation of the principle of truth condition assigned in a possible world, where it is substituted by the principle of coherence or consistency within that world. Thus, the status of a proposition P within the common ground (henceforth, CG) of a discourse can be identified as belonging to one of the following types (p. 92).

(A) P is already in the CG, or
(B) P is not in the CG, but is:
either (i) coherent with CG, in which case P is:
      either (a) a conventional assertion
      or   (b) an unconventional assertion
or   (ii) incoherent with CG, in which case P is:
      either (a) rejected as irrelevant
      or   (b) interpreted as conversational implicature and incremented as metaphorical, ironic, etc.

This is an approach to discourse as a process based on an incremental view of communication, where the definition of worlds and the rules for their acceptability and the acceptability of propositions within them is not governed by rules of logic (i.e., the rules of truth-conditional semantics or modal logic). Rather, it is governed by discourse principles, mainly, coherence and epistemic accessibility. As such, coherence is not a property inherent to the discourse itself, but it is provided by speakers and hearers, "by evaluating hypotheses formed on the basis of their knowledge of the meanings involved" (Werth, 1995b. p. 208). Truth is considered as a relative notion ranging on a scale from 100 percent certainty to 100 percent falsity, through different degrees of possibility.

With regard to knowledge and communication, Werth (1995b, p. 162) further specifies what constitutes the shared knowledge between speaker and hearer in a particular discourse (compare Givón, 1989, p. 135). Shared knowledge is divided into general knowledge and mutual knowledge, where general knowledge includes cultural and linguistic domains and mutual knowledge includes the perceptual and experiential shared domains.

3.3.2. Layering, World Builders, and Subworlds

The protagonists of the discourse situation include the participants, or interlocutors in a discourse world (speaker and hearer in a conversation, writer and reader in written communication), and the characters, which include narrators and characters within a text world (Werth, 1995b, p. 291). This distinction is based on the establishment of at least two different levels in the discourse situation, the level of the discourse world and the level of the text world.[5] Other possible layers within the text world are created by means of world building predicates, words and expressions that can create subworlds according to location (space and time deixis), modality (probability), and interaction (the use of direct speech within a narrative constitutes a subworld) (p. 283). As Werth points out (p. 286), "Sub-worlds typically use the language of what semanticists call opaque contexts to build themselves" and typically belong to the area of modality. Whereas the discourse world is mainly interactive (between speakers or reader and writer), the text world is identified deictically and displays a viewpoint—the one of a speaker-writer—for the benefit of an interlocutor; finally, subworlds are created via modal/epistemic elements that "stipulate situations which cannot (as yet) be confirmed" (p. 286).

Typical subworld-building elements are "modals, probability markers, verbs of propositional attitude, non-factive verbs, adverbials denoting imaginary, speculative or stipulative environments, and so on" (Werth, 1995b, p. 288). Some subworlds are directly accessible from the discourse world, for example those that alter the parameters of the text world by means of shifts in space or time (a flashback is a typical example). Others, however, are not directly accessible from the discourse world, as they are mediated through another entity, such as a character in the text world. Typical examples of this kind are modal worlds built by means of modalized expressions (In John's mind, Jack believed..., Mary realized..., It seemed that...). Werth (1995a, p. 77) summarizes the types of character accessible subworlds as follows.

1. Cognitive domain: in John's mind, Mary believed that..., I think, it seems..., Einstein knew..., Bill realized...
2. Intentional domain: Miriam wanted to..., in order to..., so that..., you must...
3. Representational domain: in the picture, according to Leavis, Carol dreamed that..., on TV, in the story...
4. Hypothetical domain: if..., had you not..., were you looking for...?
5. Epistemic domain: perhaps, possibly, must have, would have, certainly.

In addition to these types, Werth observes that a further type of subworld can be envisaged, the assumption, which he defines as a "proposition whose function is to help define a world rather than to denote situations which take place against the backdrop of an otherwise defined world" (Werth, 1995a, p. 78). A typical example is the *if*-clause in a conditional. Furthermore, the author stresses the crucial role played by knowledge frames in the construction of worlds because they contribute significantly to the process of "filling out and enriching the text world" (p. 78). Even if these elements are not world-builders themselves, they have more to contribute to the process of definition of the text world and subworlds than to the plot-advancing function.

The following is an example of a deictic shift in time that builds a temporal subworld (adapted from Werth, 1995b, p. 331).

(1)   The five men were spread out like the points of a five-pointed star. They had dug with their knees and hands and made mounds in front of their heads and shoulders with the dirt and piles of stones. Using this cover, they were linking the individual mounds up with stones and dirt. Joaquín, who was eighteen years old, had a steel helmet that he dug with and he passed dirt in it. He had gotten his helmet at the blowing up of the train... (Hemingway, *For Whom the Bell Tolls*)

The original paragraph is much longer but it is partially reproduced to exemplify the shift in time deixis within a text world. This takes place in the last sentence of example (1), where the change from simple past to past perfect (*had gotten*) indicates that there is a flashback. Similarly, other subworlds can be created by means of evoking action in another place (*meanwhile, back at the ranch...*), or by projecting characters' wishes (*Mary wanted to go*) or beliefs.

In brief, text worlds are defined first by specifying the deictic information regarding entities and time and space coordinates. These are the basic world-building elements. Subworlds can modify the deictic information related to these coordinates or they can introduce further domains in the text world. The three main types of subworld are summarized by Werth (1995c, p. 329) as follows.

(a)   deictic alternations: they include alternations in time and place.

(b)   propositional attitudes: they represent notions entertained by the protagonists, such as desires, beliefs and purposes.

(c)   epistemic subworlds: they are modalized propositions expressed by participants or characters. They include hypothetical worlds and modal worlds. To these the subgroup of quantity should be added: it includes quantity and negation.

World builders, such as the ones described, are differentiated from *plot-advancing or function-advancing propositions*. The difference is captured by the following two examples.

(2)   While the news was on, John finished his dinner.

(3)   While John was eating his dinner, the phone rang.
(Werth, 1995c, p. 293)

In (2) and (3), we can establish a distinction between the plot-advancing propositions *John finished his dinner* and *the phone rang*, where a certain action takes place, and world-building propositions such as *While the news was on* and *While John was eating his dinner*, which deictically contribute to the creation of a text world. Although function-advancing propositions are typical of the text world, they can also occur within a subworld, where they create parallel substories. The author provides the following example (1995c, p. 295).

(4)   At dawn today, John struggled out of bed. He took his fishing gear down to the river. If John catches a big fish, he'll take it home. He'll then give it to Mary to skin and clean. Mary will complain like hell, but she'll do it anyway. Then she'll cook it into some exotic and delicious dish.

In extract (4), the conditional *if*-clause is a world builder that projects a subworld that alters the parameters established in the text world regarding time (*at dawn today*). The subworld constitutes a kind of pause, where a hypothesis is developed about future events. It can be said that the subworld develops a (hypothetical) story-line and, consequently, has function-advancing propositions that make the story move forward within that particular world.

Depending on their nature, different text types will be characterized by different kinds of function-advancing propositions. Thus, Werth (1995c, p. 294) proposes the following classification of function-advancing propositions in relation to the four basic text types (see Table 3.1).

In the discussion of fictional writing, we are obviously interested in considering the functions of narrative and descriptive propositions. Whereas the identification of narrative is quite straightforward, "the distinction between world-building and description-advancing is sometimes difficult to draw" (Werth, 1995c, p. 305).

TABLE 3.1.

Text Type	Predicate Type	Function	Speech Act
Narrative	Action, event	Plot-advancing	Report, account
Descriptive	State, quality, habitual	Scene/Person/Routine-advancing	Describe scene/Character/Routine
Scene			
Person			
Routine			
Argumentative	Relational	Argument-advancing	Postulate, conclude, etc.
Instructive	Imperative	Goal-advancing	Request, command, etc.

TABLE 3.2.

World Building	Function-Advancing
World builders	Narration
Entity (identification)	Plot-advancing
Time	
Place	
Subworlds	Description
Participant accessible	Scene/person/habit advancing
Time	Individuation
Place	Framing
Character accessible	
Propositional attitudes	
Modalizations	

Indeed, intuitively, there seems to be a close connection between description and world building. Werth suggests that whereas world building has to do with "the presence in the text world of certain entities, including any descriptive material necessary to identify them (such as relative clauses), [description-advancing] provides further modification on elements already nominated in the text world" (p. 305). This leads Werth (p. 305) to argue that there are three basic categories of description: identifying, framing, and individuating, where "Identification is a world-building process, individuation is description-advancing (since it serves to broaden and deepen our knowledge of a nominated entity), and framing adds further information about an evoked entity from memory." The fact that the first category of description is defined as world building shows that it is not possible to draw a clear-cut line between world building and description as a function-advancing phenomenon.

The category *framing* is particularly significant when discussing negative sub-worlds, especially in negative accommodation (see Werth, 1995b, p. 198). Thus, Werth (p. 198) argues that "The frames set up the complex network of expectations which we might hold about the set of topics under discussion; the text, by way of the negative subworld, informs us that these expectations have been departed from." This can lead to the assumption that framing must be a crucial phenomenon in negation in general terms, as will be discussed in Chapter 5.

Table 3.2 provides a summary of what has been discussed so far about the world-building and function-advancing processes in text worlds.

Werth provides a clear summary of this model as follows (1995a, p. 78).

A world as we've used the term, is a conceptual domain representing a state of affairs. A text world, in particular, represents the principal state of affairs expressed in the discourse. First, the world must be defined: this is effected by means of the deictic and referential elements nominated in the text, and fleshed out from knowledge (specifically, knowledge-frames), a process I've called world building. World building, then, sets the basic parameters within which entities in the text world may operate.

In the subsequent sections, the focus is on the role of negation as a subworld in the framework under discussion.

3.3.3.   Negation as Subworld

As a propositional modality, negation belongs to the third type of subworld, that of modal shifts from the parameters set in the text world. More precisely, negation is dealt with as a form of quantification because both negation and quantifiers have to do with the question *how much?*, a scalar property (Werth, 1995c, p. 376). The difference between Werth's (pp. 373–376) categories of quantification and negation is that quantification relativizes otherwise absolute statements. Negation, however, does not alter the text world parameters by relativization but by permanently modifying previously expressed or assumed propositions. This can be observed in the examples under (5).

(5)   a.   There were eight Swedes in the room. Some were called Jan.
      b.   There were eight Swedes in the room. None was called Jan.

Whereas (5)a introduces a quantifier sub-world (by means of *some*) that relativizes the absolute nature of the preceding statement, (5)b contains a denial of an assumption that any of the men might be called Jan. The assumption may be present for whatever reason in the common ground, for example, because *Jan* is a typical Swedish name, and it creates the expectation that it should be the most common name in the set. By means of negation, this proposition, in this case an assumption, is canceled from the text world. The view of negation as quantifier complements the previously discussed view of negation as a propositional modality (see Section 2.5.2. in Chapter 2), so that negation can be seen both in terms of quantity and truth.

With regard to the scope and meaning of negation, Werth (1995c, p. 378) defends the view discussed in Chapter 2 (compare Givón, 1979, 1993, Horn, 1989) in which the interpretation of negation is dictated by the context. Thus, Werth claims (1995c, p. 378) that a de-contextualized sentence such as (6) may have at least four different interpretations.

(6)   A dog wasn't barking.

A [dog] WASN'T [barking] = It is not true that a dog was barking (widescope).
It denies the previously asserted: *A dog was barking.*

A *dog* wasn't [barking] = Not even was there a dog barking (widescope). Contrary to expectation (= *Not a dog was barking*).

A DOG wasn't [barking] = It wasn't a dog that was barking (narrowscope). Conceding that something was barking, but denying that it was a dog.

A [dog] wasn't BARKING = It wasn't barking that a dog was engaged in doing (narrowscope). Conceding that a dog was doing something, but denying that it was barking.

Werth agrees with other linguists (compare the review of Givón in Chapter 2) that negation does not consist of merely stating a negative state of events. Werth agrees with Givón (1979, Chapter 3) that negation has a foregrounding function, whereby a previously mentioned or assumed proposition is brought to the foreground and challenged in some way. According to Werth (1995c, p. 379), this is the reason why negation is rarely used to open an exchange, and it accounts for its asymmetrical relation to the affirmative. Thus, although (7)a could easily be used as discourse initial, (7)b would be less usual.

(7) a. A dog was barking.
    b. A dog wasn't barking.

Werth observes that "The essential mechanism is communicative: one does not comment on the absence of some situation unless its presence has been expected, asserted or presupposed." (1995c, p. 380). In the same line as the authors discussed in Chapter 2, Werth (p. 381) argues that a cognitive approach to the function of negation is necessary to understand its meaning in discourse. According to the author, this meaning is closely linked to the presence of expectations of some kind.

> You cannot, that is to say, negate something, unless there is a good reason to expect the reverse to be the case, whereas you *can* affirm something whether or not there is good reason to expect the opposite to be the case. The explanation for this is perfectly commonsensical: to deny the existence or presence of an entity, you have somehow got to mention it. (p. 381)

Werth, thus, takes up the asymmetricalist view of negation, which claims that its function in discourse depends on the explicit or implicit presence of a corresponding affirmative form. The relationship between the affirmative and the negative from a text world viewpoint is that "the text world identifies the common ground, or set of expectations, for the particular discourse, while a negation is naturally expressed by way of a subworld" (Werth, 1995c, p. 382). The author provides an example from *The Importance of Being Earnest*, reproduced under example (8).

(8)   ALGERNON: Please don't touch the cucumber sandwiches. They are ordered especially for Aunt Augusta.

JACK: Well, you have been eating them all the time.

ALGERNON: That is quite a different matter. She is my aunt.

LADY BRACKNELL: ... And now I'll have a cup of tea, and one of those nice cucumber sandwiches you promised me.

ALGERNON: Certainly, Aunt Augusta....Good heavens! Lane! Why are there no cucumber sandwiches? I ordered them specially.

LANE: There were no cucumbers in the market this morning, sir. I went down twice.

ALGERNON: No cucumbers!

LANE: No, sir. Not even for ready money.

(Wilde: *The Importance of Being Ernest.*)

In (8), an expectation is created for Lady Bracknell that there will be cucumber sandwiches for tea. At tea-time, however, there are no cucumber sandwiches (since Algernon has eaten them all). In terms of text world theory, this is explained as the expectation being created within the common ground of the text world, and subsequently, this information is altered by means of a negative utterance (*Why are there no cucumber sandwiches?*). This leads to a redefinition of the text world parameters, with the modification of the information that there are cucumber sandwiches. The main difference between the type of alteration of the text world parameters carried out by negation as compared to the alterations realized via other subworlds is that in the case of negation, the changes are not temporary but permanent and affect other world-building parameters. In the case of this example, Werth argues that it is the intention on the part of the character to eat cucumber sandwiches that is affected by negation, as it will not be fulfilled. The negative in a more obvious way can be interpreted as modifying the assumption/expectation that there should be a given entity—*cucumbers*—in the text world. In this sense, negation is performing a descriptive individuating function, by denying a property that was previously assumed to be applicable with regard to the entity *cucumbers*.

3.3.4.   Negative Accommodation

Although negation as the defeat of an expectation that is created within the common ground is considered to be a prototypical function of negation, Werth points out that negation can also be involved in a less prototypical phenomenon, negative accommodation (Werth, 1995c, p. 384). By accommodation, Werth refers to the phenomenon whereby entities are introduced into the discourse without being asserted explicitly. In cognitive terms, this means that new information is introduced in the discourse in a backgrounded way, typically, within dependent structures (p. 421). Thus, Werth provides the following example, reproduced in (9) (p. 404).

(9)   A: How's life with you?
     B: Great! I realized last night that my brother wasn't dead.
     A: I didn't know you had a brother!

Although new information is typically conveyed via assertions, in this case, it is less typically introduced by means of a dependent structure, the NP subject of a subordinate clause, *my brother*. The author names the phenomenon *unconventional assertion* or *accommodation*.[6]
Negation can be used in this way to deny something while at the same time presenting it and introducing it into the common ground. The author provides an interesting example from E. M. Forster's *A Passage to India*. I reproduce part of the passage under example (11).

(10)   There are no bathing steps on the river front, as the Ganges happens not to be holy there; indeed, there is no river front, and bazaars shut out the wide and shifting panorama of the stream. (Werth, 1995c, 386)

Werth (1995c) argues that the denials (*no bathing steps, not holy, no river front*) introduce the expectation that those items should be there, while at the same time they deny their presence. The expectations are culturally based and they are connected to frame-knowledge about India, rivers, cities, and so forth (Werth, 1995b, p. 196). According to Werth (1995c, p. 386), this phenomenon has two crucial consequences: (1) it is not important whether the reader is acquainted or not with the facts that are denied (i.e., that there are usually steps on river fronts in India, that the Ganges is a holy river, etc). The reader who does not know this information previously can still understand the text because the denials both present and deny the information at the same time. (2) the evocation of culturally determined frames determines the acceptability of certain sentences versus the unacceptability or oddity of others (cf. the discussion in 2.5.4. in Chapter 2). Thus, Werth (1995c, p. 386) points out that it would be odd to have a paragraph like (11) in the context of the Forster text on Chandrapore.

(11)   There are no ice-cream stands on the river front, as the Ganges happens not to be holy here; indeed, there is no river front, and restaurants shut out the wide and shifting panorama of the stream.

This is so because *ice-cream stands* and *restaurants* are not part of the frames evoked by words such as Ganges. The fact that texts such as (12) sound odd and are considered unacceptable, however, does reveal that, to some extent, it is necessary for a reader to be acquainted with the cultural knowledge that is activated in the text, contrary to what Werth seems to imply. If a reader does not have access to the cultural knowledge evoked by a text, that reader will not be able to make complete sense of the text or make a judgment on the acceptability or unacceptability of the text.

Continuing with the discussion of example (10), the process of reading involves a dynamic interaction between the different elements of the discourse and text world. Thus, the frames, which are within the text world and contain the knowledge about India, the Ganges, and so forth, together with associated inferences and connected information (e.g., antonyms), set up a complex network of expectations about the relevant topic within the text world. The relation between the processing of frame knowledge and information introduced by means of a subworld is explained by Werth as "the text, by way of the negative subworld, informs us that these expectations have been departed from" (Werth, 1995b, p. 198). In this way, a dynamic process of understanding is developed where there is a constant and continuous checking of new information against expectations created in the existing common ground.

The phenomenon of negative accommodation, attractive as it may be from a theoretical standpoint, poses problems when trying to identify it and describe it in discourse. The first difficulty is that the notion of accommodation hinges upon the problematic notion of *new information*. In Werth (1995b) new information is the information that is not recoverable from the CG of the discourse. The CG itself changes throughout the reading process, however, because it is constantly updated by the adding of new information and the deleting of old information that is no longer applicable. This should require a stricter view of *new information* as referring only to information that is not recoverable *from the immediate context of an utterance*. This allows us to consider as new, information that contains specific frame knowledge that may be part of the common ground because it belongs to higher-level frames that have already been activated or are familiar to the reader. In the case of the novel *Catch-22*, for example, the general frame for war is activated immediately in the discourse of the novel, although there are many lower-level related frames that are evoked by specific episodes. In this sense, the specific information introduced in the episodes can be defined as new if it adds details to what the reader already knows about the story of the novel and corrects assumptions that derive from the general war frame and associated expectations.

The second problem in the identification of accommodation is practical in nature, in the sense that Werth does not provide sufficient illustrations of the phenomenon. The Chandrapore episode from *A Passage to India* is the only extract the author discusses at length. The nature of the discussion leads a reader to think that the activation of frame knowledge is in some way crucial in negative accommodation, though this is not mentioned explicitly. The question then arises whether negative accommodation always involves the activation of culturally determined frame knowledge, or whether there can be cases of negative accommodation that do not necessarily require the activation of this kind of knowledge. In the discussion of extracts from *Catch-22* in this chapter and in Chapter 5, I assume that negative accommodation typically operates by foregrounding specific culturally determined frame knowledge that is then denied. This does not mean, however, that frame knowledge is not activated in the standard type of negative subworld; indeed, frame knowledge is constantly activated in discourse, as argued in Section 3.3.2.

A third problem in identifying negative accommodation is associated with the nature of negation itself. Because negation typically operates as the defeat of an expectation, what Givón defines as the *presuppositional nature* of negation, it could be claimed that negatives typically introduce the information they deny, except for those cases where the affirmative has been explicitly mentioned in previous discourse. This is obvious in negative modals, such as example (12), which is part of an episode in *Catch-22* where an officer has the idea of putting up notices postponing parades that can never take place and that have never taken place.

(12) There will be no big parade this coming Sunday.

The effect is to make people believe that Sunday parades are a usual Sunday event, however, this is not so. In this example, the effect is very similar to that of accommodation, that is, to introduce new information to deny it. For this reason, examples of negation such as this one are considered varieties of negative accommodation.

Finally, it can be argued that negative accommodation seems to be a subtype of the identification function in descriptions. As discussed in Chapter 2, new entities in a discourse are not normally defined in negative terms; for cognitive reasons, positive terms are favored. This is the reason why the identifying function in descriptive texts is less frequently performed in negative terms. In those cases, however, where we do find negative terms as new information in the discourse, we are carrying out an identifying function, which, because of the properties of negation, means the introduction of an entity to deny it, that is, accommodation.

3.3.5. The Function of Negation in Up-dating Information

Within the process of text production and understanding, the function of negation is that of up-dating information by means of altering parameters of the text world or by canceling previously held expectations.[7] According to Werth (1995c, p. 388), the information that is denied by means of a prototypical negative subworld takes the form of assumptions present in the common ground, which may derive from function-advancing propositions or which are part of world-building material, or, finally, it can derive from information that is inferred from or evoked by function-advancing propositions. Negative accommodation, however, "introduces the assumptions it is then responsible for nullifying" (p. 389).

The updating function of negation can be compared to that of rechanneling mentioned by Leinfeller (1994) and discussed in Chapter 2. As compared to the functions of other subworlds, negation in its prototypical form and in the form of negative accommodation pose problems with regard to the categorization within the class of subworlds. Thus, subworlds, in general, carry out either of two functions: (1) to temporally change parameters established in the text world (e.g., time and place shifts); or (2) to project an inaccessible state of affairs, typically by means of the expression of wishes, beliefs and intentions of characters. Werth (1995c, p. 387) observes that the first type of negation (prototypical negation) seems to strain the characteristics of the former kind of subworld, as prototypical negative subworlds constitute permanent changes in the states of affairs mentioned in the text world. The second type of negation, or negative accommodation, seems to take the notion of inaccessibility to an extreme by means of the simultaneous presentation and denial of information. The author concludes that the main function of negative subworlds is closely related to that of information up-dating or incrementation by means of topic change. Thus, although by means of other subworlds we can have temporary departures from the text world parameters, negation leads to a permanent change of a parameter. For example, a flashback does not alter the fact that the main narrative is taking place at another moment in time, and the story will at some point move from the flashback back to the main text world narrative time. Thus, the temporal parameter set up previously in the text world is not radically changed but only temporarily departed from. The same applies to the projection of subworlds by means of conditionals or world-creating predicates; in principle, these constitute only temporary shifts from the main deictic indicators of the text world. With negation, however, the departure from a parameter becomes permanent, and this leads to a change or shift in the topic or subtopic dealt with.

To sum up, negation in Werth's (1995c) text world theory is defined as a subworld that has two important discourse functions: (1) to alter or change parameters previously introduced in the common ground; and (2) through negative accommodation, to project an inaccessible state of affairs that constitutes the simultaneous presentation and denial of new information. The process by which this takes place is a dynamic process, which involves the checking of information contained in the common ground, including inferences and frame-knowledge that create expectations in the text world, and the information that is altered by means of negative subworlds. The process is one of constant up-dating and changing of the flow of information as it develops in the text world. This view of negation as an element within a dynamic discourse process is crucial for the understanding of fiction, to which we turn in the Sections under 3.4.

3.4.   AN APPLICATION OF WERTH'S MODEL TO NEGATION IN *CATCH-22*

The following sections provide illustrations of how Werth's theoretical framework can be applied to the analysis of negation in the novel *Catch-22*. The two types of negative subworlds that have been introduced, prototypical negative subworlds and negative accommodation, are discussed in turn.

3.4.1.   Negation as Subworld That Modifies Previous Information

Example (13) is an extract from a chapter devoted to a character known as Major Major. The chapter tells the story of his life, from when he was a child to the time when he is promoted as Major in the squadron on the Italian island of Pianosa, the main setting in the novel. The passage tells how Major Major, as a child, is told that his real name is different from the one he has been previously given, and that his father has deceived him and his mother by making them believe he was called Caleb. An apparently minor problem like this one turns out to have dramatic consequences on Major Major's life.

> (13) On Major Major himself the consequences were only slightly less severe. It was a harsh and stunning realization that was forced upon him at so tender an age, the realization that he was not, as he had always been led to believe, Caleb Major, but instead was some total stranger named Major Major Major about whom he knew absolutely nothing and about whom nobody else had ever heard before. What playmates he had withdrew from him and never returned, disposed, as they were, to distrust all strangers, especially one who had already deceived them by pretending to be someone they had known for years. Nobody would have anything to do with him. He began to drop things and to trip. He had a shy and hopeful manner in each new contact, and he was always disappointed. Because he needed a friend so desperately, he never found one. He grew awkwardly into a tall, strange, dreamy boy with fragile eyes and a very delicate mouth whose tentative, groping smile collapsed instantly into hurt disorder at every fresh rebuff. (p. 112)

This passage typically illustrates the function of negation of modifying assumptions and expectations that arise from the function-advancing propositions of the discourse. The extract presents a mixture of descriptive and narrative material of which the negative propositions form part. We can distinguish between two different parts in this extract regarding the projection of negative subworlds. The first part is introduced by the clause (1) *the realization that he was not ... Caleb Major*, and is expanded by two other negative clauses: (2) *about whom he knew absolutely nothing*, and (3) *and about whom nobody else had ever heard before*. The first clause denies the assumption that the character's name is so-and-so, whereas the two other clauses are relative clauses that modify the noun *Major Major*. As such, the negative relative clauses form part of what Werth defines as the descriptive function of identification: a new entity is introduced in the discourse and is here defined by means of negative properties. What is striking about this passage is that, in fact, it is not that a new entity has entered the discourse but, rather, that a property of the entity, that of having a given name, has changed. In a rather unusual process, however, the names are understood to refer to different entities, thus ignoring the pragmatic principle that makes it possible to refer to the same entity by means of different referring expressions. In this sense, it might also be argued that the contrast between the two clauses *the realization that he was not ... Caleb Major, but instead was some total stranger named Major Major Major* evokes (and denies) background knowledge in the form of logical or common sense principles regarding the identity of persons. In Chapter 5, further details are provided regarding the denial of expectations that are grounded on logical principles.

The event described in example (13) produces a strong identity crisis in Major Major, which has extreme consequences, not only on him but on the people that surround him. Seen in the light of the whole novel, where problems of identity are recurrent for several characters, we can say that this extract brings to the foreground the view of identity as something that is imposed from outside, rather than something a character develops as an internal property. Thus, it becomes more important to know a character's name (is it Caleb or Major?) than to know his internal characteristics as a person. Furthermore, the external attribute, unlike internal properties, is always arbitrary, and this arbitrariness is stressed by the fact that Major Major's father takes up the responsibility of giving his son a name as a playful joke on his fate. Thus, the clauses (2) *about whom he knew absolutely nothing* and (3) *and about whom nobody else had ever heard before*, can be said to expand on the negative subworld introduced by clause (1), by focusing on the *lack of identity* of the character determined by the fact that he has a new name.

The second part of the extract is introduced by the clause (1) *What playmates he had withdrew from him and never returned*, which introduces the second set of negative clauses[8] in this episode. In this part, the negative clauses deny assumptions and expectations that arise from the narrative material of the discourse: (2) *Nobody would have anything to do with him;* and (3) *Because he needed a friend so desperately, he never found one*. The paragraph developed by these negative clauses can be said to focus on interpersonal consequences of his change of name: people refuse to have anything to do with him any more.

To summarize, we can say that negation in this extract projects two sets of related nonfactual domains, one focusing on Major Major's loss of identity when he finds out his real name, and the other focusing of the interpersonal consequences of the first event, which lead to an attitude of rejection by others towards Major Major. Both types of subworlds modify assumptions and expectations that arise from function-advancing propositions of a descriptive or narrative type, which make reference to the properties of Major Major as a character; these changes lead to dramatic consequences. This change takes place, however, because an unusual equation is established between the identity of a person and that person's name, in such a way that the name *is* the person. In Section 3.6., I comment further on this extract by pointing out how negation is linked to the expression and development of conflict in the text world.

3.4.2. Negative Accommodation

In this section, an analysis is made of an extract where the second function of a negative subworld takes place, that of negative accommodation, as defined in Section 3.3.4. The extract closes the long episode of Clevinger's trial, when Clevinger, a clever soldier who is always trying to do his duty as a soldier well beyond what is actually required by the military regulations, is considered to be under suspicion by the higher military officers. The officers bring charges against him for no particular reason (see a discussion of this episode in Chapter 5). In example (14), Yossarian is talking to Clevinger just after the trial has finished.

> (14) Yossarian had done his best to warn him the night before. "You haven't got a chance, kid," he told him glumly. "They hate Jews." "But I'm not Jewish," answered Clevinger.
> "It will make no difference," Yossarian promised, and Yossarian was right. "They're after everybody." Clevinger recoiled from their hatred as though from a blinding light. These three men who hated him spoke his language and wore his uniform, but he saw their loveless faces set immutably into cramped, mean lines of hostility and understood instantly that nowhere in the world, not in all the fascist tanks or planes or submarines, not in the bunkers behind the machine guns or mortars or behind the blowing flame throwers, not even among all the expert gunners of the crack Hermann Goering Antiaircraft Division or among the grisly connivers in all the beer halls in Munich and everywhere else were there men who hated him more. (p. 106–107)

The episode contains the two uses of negation discussed previously, negation with a function of updating information in the text world and negative accommodation. The first function can be observed in the first part of the extract, in sentences (1) *You haven't got a chance, kid*; (2) *But I'm not Jewish*; and (3) *It will make no difference.*[9] Whereas in extract (13), the first three clauses could be seen as constituting one common domain, in this case, the interpersonal function is more prominent, and for this reason, each of the negative utterances projects a different domain that modifies the preceding utterance. The process can be described as being based on an illogical assumption: Yossarian, using one of his typical faulty logic arguments, inverts the relation *the whole can stand for the part* to establish that *a part can stand for the whole*. In the terms of the example, the higher officers are against everybody (the whole), which includes Jews (a part), consequently, it does not matter whether I take a part or the whole, the result will be the same. Yossarian's first utterance, *you haven't got a chance*, thus denies Clevinger's possible assumption that he might have a chance to be acquitted. Clevinger's reaction, *But I'm not Jewish*, denies Yossarian's implication that he might be Jewish (*They hate Jews*). Yossarian's last utterance denies Clevinger's utterance by revealing the peculiar logic of his argument, *It makes no difference*, and denying the validity of Clevinger's excuse, that of not being Jewish. The three negative utterances, thus, deny information that is inferred from preceding discourse.

From the point of view of what are standard assumptions about communicative behavior, this exchange is odd in the sense that it would have been much more economical and straightforward for Yossarian to say: *You haven't got a chance, they're after everybody*. In this way, Yossarian would be denying an assumption (and hope) that Clevinger could be declared innocent. In terms used by Relevance theory, the processing effort would be balanced by the contextual effects, as the utterance would be maximally efficient. Yossarian's utterances, however, impose a very high processing effort with apparently no gain in contextual effects, at least for Clevinger. It can be said, however, that the gain in contextual effects is directed to the reader. By expanding the exchange by means of Yossarian's inverted-logic process, attention is brought, once again, to arbitrariness as a pervading factor in life for the men on Pianosa. This time, the arbitrariness has to do with the identity of the person as belonging to a community or social group; it does not matter whether you are Jewish or not, even if Jews are pointed out as being particularly apt to be the object of persecution, thus, making reference to events in World War II. It does not matter because anybody can be persecuted, irrespective of their social identity. This seems to justify arbitrariness as a means of linguistic and logical expression. The way this arbitrariness is focused on by Yossarian's peculiar reasoning process reveals the need to understand the reading process, in general, and the function of negation, in particular, as dynamic. It is not until the reader has reached the sentence, *They're after everybody,* that the reader becomes aware of the implications of the previous exchange.[10] Thus, the reader is led from believing that the higher officers may share an underlying fascist ideology with the enemy in persecuting Jews to realizing that the higher officers' hate is extended to include everybody.

Turning now to the second part of the extract, we find that the use of negation here corresponds to what has been defined previously as negative accommodation, that is, the simultaneous presentation and denial of an item in such a way that specific frame knowledge is introduced in the discourse. This function is carried out by the following negative clauses (some of which are elliptical): (1) *nowhere in the world*; (2) *not in all the fascist tanks or planes*; (3) *or [not in] submarines*; (4) *not in the bunkers behind the machine guns*; (5) *or [not in] mortars*; (6) *or [not] behind the blowing flame throwers*; (7) *not even among all the expert gunners of the crack Hermann Goering Antiaircraft Division*; and (8) *or among the grisly connivers in all the beer halls in Munich*. The sequence of negatives can be said to constitute a discourse unit that evokes specific frames associated with the higher level frame WAR. In each clause, an entity that evokes a specific frame is presented and its presence denied: fascist tanks and planes, bunkers, machine guns, mortars, flame throwers, expert gunners, and so forth. The negative clauses can be said to project two parallel domains, one referring to the higher officers and their hatred towards Clevinger, and another one referring to the Germans, the real enemy. The presence of the Germans as enemies is precisely introduced by means of the negative clauses: *not in all the fascist tanks or planes, not even among all the expert gunners, or among the grisly connivers in all the beer halls in Munich*, and so forth. Their power as a "real" enemy, however, is denied. This leads to the identification of a potentially more fearful enemy in the higher officers of the American army (*men who hated him more*). The episode is striking precisely because it inverts the set of assumptions a soldier has regarding who is the enemy. In this passage, we are told that no matter how terrible the Germans may be, there are other men who are even more dangerous, and they are Americans too.

To summarize some of the main points discussed in this section, in examples (13) and (14), the reader's interpretation of the ongoing story is subject to continuous revision and reinterpretation as the reading proceeds. Negation carries out a crucial role in this process of up-dating and rechanneling previously held assumptions, sometimes with unexpected consequences.

3.5. POSSIBLE WORLDS AND FICTIONAL WORLDS

Possible world theory has been extremely influential with regard to the discussion of the ontological properties of fictional entities. Indeed, it presents an alternative to the speech act approach to literariness, where fictional discourse is defined as a weaker or pretended form of assertion (see McHale, 1987; Petrey, 1990; Pratt, 1977; Searle, 1975). In this view, inconsistencies within the discourse and figurative uses of language, such as metaphor and irony, are interpreted by means of indirect speech acts or by means of conversational implicature (see Pratt, 1977). Whereas speech act theory concentrates on the characteristics and the definition of the type of force of the utterances expressed within the fictional world, and of the fictional text itself as a *mega utterance*, possible world theory applied to the study of fiction focuses on the definition of the boundaries of a fictional world and the description of the internal properties of that world. As Doležel (1989, p. 221) observes, possible world theory considers fictionality from the perspective of two problems: (1) the ontological status of fiction as a nonexistent object; and (2) the logical status of representation. Doležel (1989, pp. 228–229) points out that, while in pragmatic theories these problems are solved in terms of conventions and pretended assertions as explained previously, possible world theory offers a view based on two main principles (Doležel, 1989, pp. 228–229.).

1. Possible worlds are possible states of affairs.
2. The set of fictional worlds is unlimited and maximally varied.

The former principle solves the problem of assigning reference to nonexistent entities, as reference will take place within the particular world where the entity is located. For example, Hamlet, as a character in a fictional play, has sense and reference within that play, within that world. The latter principle allows for a redefinition of possible world theory where even a world that does not follow the Law of Non-Contradiction is possible. This view is based on the assumption that even the "real" or actual world that we inhabit is a construct (see also Eco, 1989; Lewis, 1979; Ryan, 1991b); this means that it is also defined as a possible world among others. Each world is the actual world for the characters who inhabit it (see Lewis, 1979, p. 184). This assumption has important consequences on the way we conceptualize the relation between fiction and reality. As McHale points out (1987, p. 34), the possible worlds approach both complicates the view of the internal ontological structure of fiction and it also blurs the external boundary that differentiates it from reality. According to McHale (1987, p. 34) classical pragmatic and semantic theories have been particularly careful to keep the distinction of boundaries between reality and fiction as clear as possible. Possible world theory as outlined by Lewis (1979), Doležel (1989), and Ryan (1991b) leads to a weakening of the limits between different worlds, bringing into focus aspects where fiction and reality overlap and diverge.

Indeed, according to Eco (1989, p. 344), the functional interest of possible world theory applied to the understanding of fiction lies precisely in its adequacy for the explanation of how a possible world diverges from the actual world. Possible world theory taken in these terms seems to be particularly adequate for the description of conflictive texts from an ontological perspective, an aspect that is typical of postmodernist fiction. For a discussion of how conflictive ontologies are accounted for in terms of this theory see McHale (1987) and Doležel (1989).

3.5.1.   Ryan's Model of Fictional Worlds

The following sections are devoted to a discussion of Ryan's (1985, 1991a, 1991b) contribution to possible world theory as applied to the description of conflict in fictional worlds. As explained previously, this framework is intended to complement what has been said about text worlds in the previous sections, by providing further instruments of analysis that focus on conflict and its expression in the fictional world.

Ryan takes up the view of possible worlds where a fictional world is seen as a possible world, and where reality, or our actual world, is also another possibility within what is better defined as a universe of alternate possible worlds. Ryan (1991a, p. 553) points out, however, that the acceptance of the ontological status of fictional texts as possible worlds is not enough in itself, and it certainly does not justify the existence of fictional worlds that are internally contradictory or otherwise conflictive. Further specifications regarding the notion of possibility and the accessibility relations between different worlds need to be made to provide a convincing account of fiction in terms of possible world theory.

Ryan (1991a) takes up Kripke's (1971, p. 64) original distinction of the three basic elements regarding the notion of a possible world and also borrows the important notion of accessibility, which substitutes that of logical possibility, as explained in section 3.2. According to Kripke (1971) a possible world is defined according to three main elements: the actual world (henceforth AW), the possible world (henceforth PW), and the relation between them (R). On the basis of these observations and following Doležel's theory of multiple worlds, Ryan (1991a, p. 554) develops a model based on the idea that a text as a semantic domain is not a single world but, rather, it projects a "system of worlds" (1991, p.) or universe that is centered around the *textual actual world*. Further, the mental representations produced by characters' beliefs, wishes, dreams, fantasies, and so forth, constitute alternate possible worlds within the textual system (cf. Werth's system of subworlds described in Section 3.3.2.).

Ryan's (1991b) theory provides a combination of Lewis's (1979) indexical theory of possibility, where "every possible world is real" (Ryan, 1991b, p. 18) and Rescher's (1979) defense of the privileged status of reality as "the actual world." Ryan does this by means of the notion of *recentering*, whereby a possible world becomes temporarily the actual world, as in dreams, hallucinations, and children's make-believe games. In these cases, there is a temporary shift from the parameters defined in the actual world to those established in an alternate possible world. According to this, possible worlds are actual worlds from the point of view of their inhabitants (Ryan, 1991a, p. 554). The notion of recentering also allows us to reconcile the view that there is only one actual world, in the sense of "real world," and the fact that fictional worlds, which in absolute terms are alternate possible worlds, are accepted "as if" they were temporarily actual worlds themselves.

Ryan (1991a; 555) observes that the notion of fictional recentering is based on a distinction among three modal systems: (1) the actual world (AW); (2) the textual universe, at the center of which is the text actual world (TAW); and (3) the text reference world (TRW), the system the text actual world represents. Ryan further explores the relations between the three modal systems and, in particular, the degrees of divorce that may take place between them. These differences revolve mainly around the factors that distinguish the AW and the TAW or TRW in different genres. In fiction, which is our main concern here, the divorce lies in the fact that the TAW does not refer to the AW but to the TRW.

> Fiction is characterized by the open gesture of recentering, through which an APW (Alternate Possible World) is placed at the center of the conceptual universe. This alternate possible world becomes the world of reference. The world-image produced by the text differs from the AW—except in the genre of true fiction ... but it accurately reflects its own world of reference, the TRW, since the TRW does not exist independently of its own representation. The TAW thus becomes indistinguishable from its own referent. (Ryan, 1991a, p. 556)

The differences between AW and TAW are further accounted for in terms of accessibility relations between worlds, which is discussed in Section 3.5.1.1.

3.5.1.1. Accessibility relations between worlds

A world is defined as possible when it is accessible from the world at the center of the system. In philosophy and logic, possibility and hence, accessibility, are interpreted in logical terms, so that a world is possible if it satisfies the Laws of Non-Contradiction and of the Excluded Middle. As explained in Chapter 2, this means that in order for a proposition to be acceptable in a possible world, it has to be either true or false in that possible world but not both. This view would immediately classify many fictional worlds as impossible worlds. Ryan (1991a, 1991b), following other philosophers such as Kripke (1971) and Lewis (1979), proposes a redefinition of the notion of accessibility that is based on epistemic rather than logical principles, and that establishes different criteria, according to which a given world can diverge from the characteristics of our actual world as we know it.

With regard to this point, Ryan (1991a, p. 558) establishes a difference between two types of transworld relations.

(a) the domain of the relations between AW and TAW, which is a trans-world or trans-universe relation;
(b) the intra-universe domain of the relations which can be established between the TAW and the alternatives that are projected within that world.

The relations between AW and TAW describe the degree of resemblance (and divergence) between the system represented by the TAW as compared to reality. The relations between worlds within a system provide the internal structure of the textual universe (Ryan, 1991a, p. 558). The former is discussed in this section. The latter is discussed in section 3.5.1.2.

Ryan (1991a, pp. 558–559) provides a list of what she considers to be the accessibility relations from the AW that are involved in the construction of the TAW. They are reproduced in the following list.

1. *Identity of properties.* The TAW is accessible from the AW if the objects common to TAW and AW share the same properties.
2. *Identity of inventory.* The accessibility depends on whether the TAW and the AW are furnished with the same objects.
3. *Compatibility of inventory.* The TAW is accessible from the AW if the TAW's inventory includes all the members of the AW, as well as some native members.
4. *Chronological compatibility.* The TAW is accessible from AW if it takes no temporal relocation for a member of AW to contemplate the entire history of TAW.
5. *Physical compatibility.* TAW is accessible from AW if they share the same natural laws.
6. *Taxonomic compatibility.* TAW is accessible from AW if both worlds contain the same species and they have the same properties.

7. *Logical compatibility.* TAW is accessible from AW if it follows the Laws of Non-Contradiction and of the Excluded Middle.
8. *Analytical compatibility.* TAW is accessible from AW if they share analytical truths, that is, if objects designated by the same words have the same essential properties.
9. *Linguistic compatibility.* TAW is accessible from AW if the language in which TAW is described can be understood in AW.

Ryan (1991a) points out that a TAW may be accessible from the AW regarding all aspects but there is always one in which they differ, and that is the fact that the sender of the fictional text is situated in the AW and not in the TAW or TRW. According to Ryan (p. 33) identity of properties and of inventory are characteristic of accurate nonfiction, such as journalism, and true fiction, such as stories based on true facts, such as, for example, Truman Capote's *In Cold Blood*. In historical novels like *War and Peace*, on the other hand, we have identity of properties, but not identity of inventory, where this is substituted by "expanded inventory" (p. 33). In the genre referred to as "historical fabulation" (p. 36), identity of properties is transgressed to create alternative possibilities to the lives of historical characters. This would be characteristic of a story where Napoleon travels to New Orleans or Hitler wins the war. Emancipation from chronological compatibility is characteristic of science fiction, which might otherwise maintain all other accessibility relations intact. Transgression of natural laws is found in fictions about ghosts or Kafka's *Metamorphosis*, and transgression of taxonomic compatibility in fairy tales with TAWs populated by fairies and dragons, for example, *The Lord of the Rings*. Emancipation from logical compatibility leads to nonsense, such as nonsense poetry or postmodern fictions where contradictory events are part of the fiction. Nonsense or absurdity can also be induced by transgressing analytical compatibility (p. 38) in such a way that objects described do not have the properties they have in the AW, for example, a horse that has the properties of a computer. To end, linguistic incompatibility appears in fictions that make use of invented language, such as Lewis Carroll's *Jabberwocky*. Ryan (p. 39) also points out that there may be undecidable relations, which arise from internal instabilities of the text world.

In the terms described previously, the TAW of *Catch-22* can be said to conform to identity of properties, expanded inventory, physical compatibility, taxonomic compatibility, analytical compatibility, and linguistic compatibility. This has to do with the pseudo-historical character of the novel, which is supposed to tell the story of an American squadron in Europe during World War II. It does not conform to compatibility of inventory, however, as imaginary characters populate the fictional world. It does not conform to chronological or logical compatibility either. These last two are closely connected. The transgression of chronology is manifested in the chaotic structure of the novel, whose time development cannot be recovered linearly without incurring in contradiction (see the discussion in Chapter 1). Although these contradictions are not immediately obvious to the reader, and thus prevent the novel from being truly postmodern in this sense,[11] the transgression of temporal sequencing does contribute to the projection of a world that seems to be suspended in time and where it is difficult to establish a beginning and an end. This is also related to the images of circularity and closedness that characterize this fictional world. These aspects are reinforced by the emancipation from criterion 7. *logical compatibility*. Ryan observes (1991a, p. 565):

> Emancipation from [7.] Logical, opens the gates to the realm of nonsense....nonsense is characterized by the rejection of the law of noncontradiction. *P* and *not-P* can be true, not just in separate worlds of the textual universe, but in its actual world as well.

It is important to point out that the transgression of the logical component in *Catch-22* does not take place in the way that it does in postmodern fiction, where this kind of transgression is manifested by means of contradictory events or states, such as the fact that a character can be both dead and alive at the same time or that alternative story lines take place simultaneously. In *Catch-22*, the transgression of logical principles is carried out by means of linguistic manipulation. It is the language use that is not logical. Ryan does not consider this possibility, as the only linguistic transgressions she mentions are the use of nonsense language or the use of an invented language. In *Catch-22*, logical incompatibility is manifested by means of a kind of determinism that establishes a cause–effect relation between language use and reality. Linguistically, this can be described in general terms as a violation of pragmatic principles rooted in logical relations.

An example is that of extract (13), discussed in the last sections of this chapter, where we are told how Major Major realizes that his name is different from what he has thought for several years. The realization that he has a different name produces an identity crisis in him, which changes his life radically as nobody else recognizes him for who he is any longer. This can be explained as a phenomenon where language determines reality in a way that goes against standard assumptions about logical relations and how they are coded pragmatically. Thus, the logic of the world of *Catch-22*, as manifested in the episode about Major Major, seems to rely on the assumption that there is a one-to-one relation between a name and a referent. That is, if the name *Caleb* stands for a particular entity in a world, the name *Major* must stand for a different entity in the same world. This reasoning process goes against the well-known pragmatic principle that different referring expressions can be used to pick out one single entity at a time (e.g., I can refer to a certain person as *that man over there, Mr. Smith*, or *the philosophy professor*).

Whereas the set of criteria discussed in this section identify differences and similarities between the fictional world and the actual world, the categories discussed in Section 3.5.1.2. provide the means of analyzing degrees and types of conflict within the fictional universe.[12]

3.5.1.2. The narrative universe

The idea of how conflict develops within the fictional universe is dealt with by Ryan in an (1985) article and in Chapter 6 of her (1991b) book. In these two works, an initial distinction is established between two types of worlds, which correspond approximately to the definitions of "text world" and "subworld" in Werth's (1995c) framework, and their corresponding relations of accessibility. In Ryan's work (1985, p. 720), we have a distinction between the text actual world, which has "an autonomous or absolute existence" within the universe where it belongs, and characters' domains, "whose existence is relative to somebody, i.e., which exist through the mental act of a character" (Ryan, p. 720). Ryan (1991b, p. 112) also establishes a difference of interpretation levels, which can be compared to Werth's distinction between the *discourse world* and the *text world* (see Sections under 3.3). In Ryan's work (p. 112), the distinction is established as a difference between what she calls the *semantic domain* of the text, which contains "all the meanings suggested by a text, the set of all the valid inferences and interpretations," and the *narrative universe*, "the collection of facts established for the various worlds of the system." This distinction seems to require the presence of the reader in the concept of *semantic domain*; whereas the narrative universe is defined as existing independently from reader and writer. The definitions used by Werth seem more adequate for the explanation of the textual and contextual components of the two notions, in particular with regard to the degree of detail in accounting for the communication process from a more rigorous linguistic perspective, however, Ryan's concepts are valid for her own purposes. For this reason, the details of Ryan's (1991b) definitions regarding the textual universe as an ontological entity are not discussed, however, in the next chapters in this book, I adopt Werth's more linguistically oriented definitions of text and discourse.

3.5.1.3. The components of the fictional world

In Ryan's work (1991b), the text actual world (TAW) of the fictional universe is defined as "a succession of different states and events which together form a history" (p. 113). In addition to this central domain, there are other regions that can exist absolutely or relative to the private world-view of specific characters (cf. Werth's participant-accessible and character-accessible subworlds). For example, in the book *Alice in Wonderland*, there is a clear difference between the TAW, where Alice and her sister are sitting on a bank, and the alternate world created by Alice by means of her dream. In this case, the dream occupies the center of the story. In other fictions, it can be otherwise. Strictly speaking, all subworlds constitute alternate possible worlds with respect to the central TAW. Ryan (p. 114) distinguishes the following types.

1. **Authentic worlds:** They are the opposite of pretended worlds, and imply that the propositions creating them are sincere. There are several types.

(a) **K-world** (or belief/knowledge world): Each K-world is defined with regard to the realization of relevant propositions with reference to the operators of necessity, possibility, or impossibility reinterpreted in narrative terms. Thus, in the epistemic system of knowledge, they are translated into knowledge, belief, and ignorance (Ryan, 1991b, p. 114). Ryan (p. 115) points out that the concept of knowledge is straightforward, and is described as "a character 'knows' a proposition p, when he or she holds it for true in the reference world and p is objectively true in this world." The notion of impossibility would involve a contrary-to-fact proposition. Possibility involves an incomplete representation, and is more complex and more difficult to account for. The most interesting observation with regard to this type of K-world is its characteristic indeterminacy, which may take place for one of two reasons: incompleteness or partiality. An incomplete K-world leaves propositions unanswered. Ryan provides an example of a detective wondering who might be the murderer in the case he's working on. A partial K-world involves not knowing some of the facts from the actual world. Ryan gives an example of a character who is unaware that another character has been murdered. According to this, we can establish a system for the computation of characters' K-worlds by means of checking propositions by assigning them the following operators (p. 115).

+   (correspondence, knowledge): x holds p firmly for true
−   (conflict, misbelief): x holds p firmly for false, while p is true
    (absence, ignorance): p is unknown to x
i   (indeterminacy, uncertainty): x is either uncommitted to the truth of p or leans to some degree toward the truth (i.e., considers p possible, probable, unlikely, etc.)

Finally, Ryan (p. 116) points out that K-worlds may contain other K-worlds, so that a potentially infinite structure of embedded K-worlds can be created. Hypothetical structures are classified as prospective extensions of K-worlds and are often related to the projection of plans and goals.

(b) **O-world** (or obligation world): This type of subworld is defined as "a system of commitments and prohibitions defined by social rules and moral principles" (Ryan, 1991b, p. 116). These rules allow us to classify actions as allowed (possible), obligatory (necessary), or prohibited (impossible). Stability and lack of conflict with respect to this subworld are defined by the realization of all obligations and the nonrealization of transgressions. Variations of this pattern lead to the establishment of different conflictive plots and different resolutions, such as the punishment of infringement of laws or the reward of the realization of merits. Other conflicts can arise because characters may belong to different communities with conflicting obligations.

(c) **W-world** (or wish world): This world establishes what is good and what is bad in terms of the individual's desires. Thus, it contrasts with the moral obligations institutionally imposed by means of O-worlds. Typical desired states are the possession of objects or other desired things, such as success, wealth, and so forth. Typical desired events are gratifying activities (Ryan, 1991b, p. 117). The satisfaction of a W-world takes place when the propositions labeled as "good" are satisfied in the TAW. The nonsatisfaction of desires will produce conflicts of different types. An important observation made by Ryan is that the notions of "good" and "bad" are relative and should be seen as extremes on a cline that can vary throughout the fictional work.

2. **Pretended worlds:** Unlike authentic worlds, they involve the entertainment of insincere propositions. A character may construct a private world in order to deceive another character. In this way, the complete domain of a character will contain sincere beliefs, obligations and wishes, and mock beliefs, obligations and wishes if he or she constructs pretended worlds.

3. **F-universes:** In addition to the subworlds described, there is another type labeled as F-universes, which express creations of the mind, such as dreams, hallucinations, fantasies, and fictional stories created by characters within the fictional world. They are defined as universes because they require an operation of recentering, whereby actuality is temporarily shifted to what in global terms is seen as an alternate possible world. Thus, a dream will have an actual domain and a set of characters' domains and subworlds, just in the same way as the text actual world.

3.5.1.4. Conflict in the fictional world

As in Werth's work (1995c), Ryan's (1991b) model of fiction is a dynamic system, where such dynamism is dictated by the desire of characters "to make the TAW coincide with as many as possible of their private worlds" (p. 119). In this view, conflict develops when there is an imbalance between the private domain of a character and the status quo of the TAW. Particularly important is the observation that conflict is not just a factor in the complication of plot but, rather, a more or less permanent condition of fictional worlds. This observation has always been valid for drama, and it can be argued that in narrative, conflict is also a primary source for the development of a story (arguably, conflict is a permanent state of ordinary life).

Ryan (1991b, p. 120) establishes a typology of narrative conflicts depending on (1) the type of subworld that departs from the TAW; and (2) "the relative position of worlds within a character's domain."

1. **Conflicts between the TAW and private worlds:**   The most frequent conflict of this type involves an unrealized wish. This pattern gives rise to the quest, which is typical of medieval literature and fairy tales but is also present in a less obvious way in other genres. A conflict with an O-world leads to typical patterns of prohibition–violation–punishment, mission–accomplishment–reward, and so forth. A conflict involving K-worlds can produce errors, which can be spontaneous, as in tragedy, or the product of deceit, as in comedy. It can also produce enigmas, which correspond to indeterminate or incomplete K-worlds (Ryan, 1991b, p. 121).

2. **Conflicts within a character's domain:**   Conflicts of this type arise when the satisfaction of one world within a character's domain implies the nonsatisfaction of another (Ryan, 1991b, p. 121). They may include patterns such as a conflict between a W-world and an O-world, when social conventions prohibit behavior desirable for characters, such as murder or adultery.

3. **Conflicts within a private world:**   In this case, private worlds cannot be satisfied because there are internal inconsistencies within worlds, such as contradictory desires or a lack of definition of the borders of internal worlds, as in postmodern fiction (Ryan, 1991b, p. 122).

4. **Conflicts between the private domains of different characters:**   This type of conflict is possibly the most usual source of conflict in narrative because it is based on the creation of antagonism between different characters. Such antagonism is rooted in the fact that the satisfaction of the private world of one character will imply the nonsatisfaction of a private world in another character. This is most obvious in cases where two (or more) characters are opposed in a hero–villain pattern (Ryan, 1991b, p. 122).

3.6.   AN APPLICATION OF RYAN'S MODEL TO THE ANALYSIS OF NEGATION AND CONFLICT IN *CATCH-22*

In this section, Ryan's (1991b) framework to the analysis of conflict is applied to an extract from *Catch-22*. The extract is the same one discussed in Section 3.4.1., where it was analyzed from the perspective of Werth's (1995c) framework. The extract is reproduced under example (15) (for an explanation of the context in which it occurs see Section 3.5.1.).

(13) On Major Major himself the consequences were only slightly less severe. It was a harsh and stunning realization that was forced upon him at so tender an age, the realization that he was not, as he had always been led to believe, Caleb Major, but instead was some total stranger named Major Major Major about whom he knew absolutely nothing and about whom nobody else had ever heard before. What playmates he had withdrew from him and never returned, disposed, as they were, to distrust all strangers, especially one who had already deceived them by

pretending to be someone they had known for years. Nobody would have anything to do with him. He began to drop things and to trip. He had a shy and hopeful manner in each new contact, and he was always disappointed. Because he needed a friend so desperately, he never found one. He grew awkwardly into a tall, strange, dreamy boy with fragile eyes and a very delicate mouth whose tentative, groping smile collapsed instantly into hurt disorder at every fresh rebuff. (p. 112)

As in Section 3.4.1., we can consider the extract as consisting of two different parts, each projecting complementary domains. The first domain is introduced by the sentence *It was a harsh and stunning realization that he was not, as he had always been led to believe*, and which goes on up to *about whom nobody else had heard before.* The word *realization* is a world- creating predicate, which introduces the domain expanded by means of the *that*-clause. The negative subworld (*that he was not...*) provides evidence of the existence of a conflict between Major Major's K-world, or knowledge/belief world, and the status quo in the text actual world. This can be expressed as meaning that Major Major thinks he is somebody and finds out that he is somebody else. The conflict arises because of the character's ignorance of the facts in the actual world.

The second part of the extract, which goes from *What playmates he had withdrew from him* to the end, expresses a different conflict, which is the direct consequence of the previous event. This conflict is between the different domains of Major Major and other characters. The conflict can be described as a conflict between different wish-worlds, where Major Major's wish to have friends and be accepted socially is not satisfied, as it is contrary to other characters' wish worlds to have nothing to do with him. Although the reaction on the part of the characters towards Major Major may seem unjustified, it shows a recurrent aspect in the novel, which is the development of conflict from a situation of deceit. In this case, the deceitful situation has been created by Major Major's father, who has made his son believe he was called Caleb. This has produced an imbalance in the text world, between the actual situation (his son's name is Major) and the knowledge/belief worlds of his son and everybody else (his name is Caleb).

A striking feature in this episode from this perspective is that such a trivial question as a character's change of name can lead to a conflictive situation and to the marginalization of the character. In addition to the idea of arbitrariness mentioned in Section 3.4.1., referring to the way in which identity is identified by means of external attributes such as names, we can also talk of absurdity in that characters rely more on external temporary attributes rather than on internal permanent properties.

Now, if we consider the type of complex world projected in this extract, the question may arise to what extent it conforms with the characteristics of the real world and to what extent it differs from it by using Ryan's (1991a, pp. 558–559) table of accessibility relations between worlds to identify such similarities and differences (see Section 3.4.1.1.). As has already been pointed out, this extract shows a transgression of logical principles, in the sense that an unacceptable cause–effect relation is established between a linguistic act of naming and a "real" act of being: you are your name. Furthermore, I have also pointed out that there was a violation of a pragmatic principle whereby one referring expression can be used to pick out different entities in the world. The combination of these transgressions with otherwise normal features produces the odd image of this world. More information is needed, however, to identify in a more exact way why it is that a description such as the one discussed here is unacceptable for the reader. This is expanded upon by making reference to how knowledge is stored and processed and how expectations are defeated in the reading process, and is discussed in Chapter 4.

3.7.   CONCLUSION

This chapter focuses on a discussion of two approaches to text world theory that have expanded on the approaches to negation as a discourse phenomenon discussed in Chapter 2. I argue that Werth's (1995c) discourse framework can contribute to the interpretation of the functions of negation within the perspective of a dynamic discourse framework by systematizing the cognitive properties of negatives in discourse.

Werth's model is complemented by incorporating the typology of accessibility relations and conflict types proposed by Ryan (1991b) in her work on narrative fiction. Within this model, it can be argued that negation plays a crucial role in the manifestation of transgressions of logical principles within the fictional world of *Catch-22*, by means of contradictions and the establishment of unacceptable logical relations of cause–effect and reference. It is necessary, however, to pursue a deeper understanding of the way in which such incompatibilities arise; thus, in Chapter 4, I argue that schema theory can contribute to the present chapter by providing the means of analyzing how expectations are created and defeated in the reading process.

NOTES

1.   See Enkvist (1989), Partee (1989), and Petöfi (1989) for different approaches to the integration of possible world theory into linguistic theory. Enkvist (1989) concentrates on the integration of possible world semantics in the process of comprehension and understanding of text. Partee (1989) focuses on the relevance of possible world theory to propositional semantics. Petöfi (1989) summarizes some of the main papers collected by Allén (1989) considering the contributions of possible world theory to linguistics in more

general terms, though focusing on semantic and pragmatic aspects related to the sense–reference distinction.

2.   Werth (1995a, 1995b, 1995c) distinguishes between two types of accessibility relations, which depend on the epistemic accessibility on the part of the reader with respect to domains defined within the text world. Thus, if a shift in time or place is presented directly by the narrator, these shifts from the basic text world story line will constitute domains directly accessible to the reader. They receive the name of *participant accessible subworlds*. If a shift from the text world parameters is introduced by means of a projection from the mind of a character, however, such as a wish, a hypothesis, a belief, the domain defined hereby will not be directly accessible to the reader, as the reader has access to such a type of domain indirectly. These receive the name of *character accessible subworlds*.

3.   For a discussion of the ontological status of possible worlds and their relation to the actual world see Loux (1979).

4.   For a discussion of the notion of frame as contrasted to other schema theoretic frameworks, see Chapter 4.

5.   See also Leech and Short's (1981, p. 281) classification of the different levels of interaction in the reading process.

Writer–Reader
Narrator–Interlocutor
Character–Character

6.   The notion of *unconventional assertion*, of course, provides a different approach to the traditional notion of presupposition. According to Werth (1995c, p. 395) presuppositions are "completely dependent upon the context they occur in," and not all the cases which are traditionally treated as presuppositions are considered to be so from the view of unconventional assertion. Thus, those cases where new information is introduced in a backgrounded way, as the examples discussed in the present section, are not considered to be presuppositions at all, but a type of unconventional assertion.

7.   By world parameters, Werth (1995c, p. 296) refers to deictic or world-building elements, and knowledge frames together with associations and inferences.

8.   In Chapter 5, I claim that negative subworlds can be defined either as sentence units or as discourse units, depending on whether the scope of the negative is understood to stretch beyond the boundaries of the sentence where it appears or not.

9.   Negation of modal verbs and other world-creating predicates are discussed in Chapter 5.

10.   My attention was directed to this fact by Clara Calvo (personal communication).

11.   See McHale's (1987) discussion of postmodern fiction as compared to modernist fiction, and Aguirre's (1991) outline of postmodern features, in particular, the violation of the law of non-contradiction.

12.   For a review of Ryan's notion of fictionality based on these notions see Semino (1993). Semino criticizes Ryan for her rigid definition of fictionality as based on expectations from real world facts.

4

Negation, Frame Semantics, and Schema Theory

His specialty was alfalfa, and he made a good thing out of not growing any. The government paid him well for every bushel of alfalfa he did not grow. The more alfalfa he did not grow, the more money the government paid him, and he spent every penny he didn't earn on new land to increase the amount of alfalfa he did not produce.

(Heller, 1961, p. 110)

4.1. INTRODUCTION

This chapter discusses the contributions made by frame semantic theory to the understanding of negation and of schema theories to the conception of discourse as a dynamic phenomenon. The frameworks discussed in this chapter overlap, to some extent, with the text world approaches discussed in Chapter 3, in the foregrounding of the cognitive aspects of text processing and understanding. In fact, a text world model like the one proposed by Werth (1995c/1999) and other authors, which contains the notion of frames, can easily incorporate certain principles of text processing and understanding that are based on schematic knowledge and how it is organized and activated during reading. The possibility of combining text world theory and schema theory is discussed at length by Semino (1994, 1995, 1997) in an analysis of text worlds in poetry, where she proposes a combination of text world principles in the description of the internal structure of fictional worlds and schema theoretic principles applied to the processing of information projected by the fictional worlds (see the discussion of Ryan's [1991b] framework in Chapter 3). Semino (1994, p. 125) argues that schema theory can contribute to text world theory because

> the reader's perception of the world projected by a text depends on the way in which his or her background knowledge is reinforced or challenged during the process of interpretation.

Frame semantics and schema theoretic models can contribute greatly to the understanding of negation in discourse. The asymmetrical relation of negation with respect to the affirmative, and its description as involving the defeat of an expectation, as discussed in Chapter 2, makes this structure particularly accessible to a schema theoretic approach. Schemata are standardly defined as expectations; if negation is understood as the defeat of an expectation, we can understand the relation between a negative and a positive term in terms of the relation between the schemata or frames evoked by each term.

4.2. SCHEMA THEORETIC APPROACHES TO TEXT PROCESSING

In the following sections, we consider the relevance of schema theoretic principles and their application to the understanding of negation.

4.2.1. Historical Background

The notion of schema goes back as far as the philosopher Kant, but is usually understood as developing from Bartlett's (1932) work *Remembering*, where he set the basis for later schema theoretic models in the 1970s. Bartlett's investigations showed that human memory of perceptual and textual data works by means of reference to previously lived experiences, which are activated during the recall process. This accounts for differences in recall of the same textual or visual data by different people, and, more particularly, for difficulties in remembering and understanding data from a different culture. Thus, the author carries out experiments where a story from a Native American community in North America is read to American participants, who are then asked to reproduce it. In the reproduction, it was proved that individuals changed the original story to adapt it to their own experience, thus providing evidence for the existence of schematic packages of information according to which we interpret new experience. Bartlett (1932, p. 12) points out that it is necessary to develop a theory that accounts for how new schemata are produced, since schemata are not static elements but units with which we do things. This is a crucial question in the theories developed in the 1970s, when a revival of schema theory took place.

Psychologists such as Bateson and Minsky and social anthropologists like Goffman contributed greatly to the development of schema theory in the 1970s. Bateson (1972) developed the idea of a *frame* as a metalinguistic signal on the basis of well-known research on the behavior of monkeys. According to this author, the monkeys he observed used some kind of signal to tell a partner whether the actions they were engaged in were "serious," like fight, or "playful." These observations were applied successfully to the observation of human activities, including the engagement in games of make believe and the creation of fictional situations (see the discussion of this theory with reference to the notion of fictionality in Chapter 1). Goffman (1974) applies the notion of frame to the classification of different interactive situations in communication. Minsky (1975) defines *frames* as units that represent stereotyped situations with attached information regarding the expectations we have of that particular situation and our behavior in it. Minsky also introduces the notion of *slots* that are filled in at particular situations. Thus, as illustrated by van Dijk and Kinstch (1983, p. 47), we can have a BUS schema that contains a series of variables, such as the actor roles *driver* and *passenger*. These variables are replaced by constants in specific situations, where they refer to particular persons. This view is the one that is standardly referred to when talking about frames and schemata in general terms.

A major contribution to the development of schema theory has been that of researchers in the fields of artificial intelligence (Rumelhart, 1980; Schank, 1982; Schank & Abelson, 1977) and second language acquisition (see, for example, Carrell, Devine, & Eskey, 1988). Schank and Abelson developed the well-known model where they systematize schemata according to four main types: scripts, plans, goals, and themes. The dynamic aspects of processing that were not dealt with in this work are the focus of Schank's *Dynamic Memory*, where the author develops a model that can account for how schemata change and how new schemata can be created. This issue, also dealt with by Rumelhart, is discussed more in detail in Sections 4.2.5. and 4.3.2.

Schemata are also incorporated in text or discourse theories as part of a cognitive theoretical background that accounts for aspects of text processing and understanding in the framework. This is the case of the discourse theories developed by van Dijk and Kinstch (1983), de Beaugrande (1980), de Beaugrande and Dressler (1981), and Brown and Yule (1983), where schemata are considered to play crucial roles in the way text coherence is understood. Finally, schemata or frames also play significant roles in the works of cognitive linguists such as Langacker (1987) and Lakoff (1989).

A problem with theories based on schema theoretic principles is the lack of agreement on terminology issues. Thus, although the term *schema* has been used in general terms by artificial intelligence researchers (see Schank & Abelson, 1977) and English as a Second Language (ESL) acquisition theories (see Carrel, Devine, & Eskey, 1988), other terms denoting similar concepts have been used in other fields. Examples include *frame* (see Bateson, 1972; Fillmore, 1985; Goffman, 1974;

Minsky, 1975; Werth, 1995c/1999), *script* (Schank & Abelson, 1977), and *scenario* (for a review of the terms see Brown & Yule, 1983, Chapter 4). For example, Schank and Abelson (1977) first introduce the now popular term script to refer to frames that contain the conceptualizations of complex situations, such as the RESTAURANT script.

The notion of frame is also incorporated by Werth (1995c/1999) in a text world model. In Chapter 3, we saw that frames were introduced in this model together with world-building predicates as significant factors that contribute to the creation and fleshing out of a text world. Werth (1995b, p. 197) distinguishes between frames that are derived from culturally shared expectations and other frames, which are specific to individual experiences. Furthermore, Werth argues that frame knowledge can be learned in two different ways, (1) by direct experience; or (2) by inferencing. Learning frames by inferencing often takes place by linguistic means, as we saw in the examples of negative accommodation in Chapter 3, where an item was intro-duced and denied at the same time. Frames in Werth's model, then, constitute packages of information, both idiosyncratic and culturally shared, which enable the reader to fill out the details regarding the definition and description of the text world. They contain "general knowledge and connected information, including opposites (antonyms)" (p. 198), and they determine the expectations a reader will develop throughout the process of reading a text.

I use both the terms schema and frame to stand for the same kind of concept, that is, a packaged unit of information activated in discourse. It can be simple or more complex, in which case the specific type of schema or frame referred to is defined at the appropriate places.

4.2.2. Characteristics and Functions of Schemata

Schemata and frames are defined in the following ways by different authors.

> A schema then, is a data structure for representing the generic concepts stored in memory. These are schemata representing our knowledge about all concepts: those underlying objects, situations, events, sequences of events, actions and sequences of actions. (Rumelhart, 1980, p. 33)
>
> A schema is a description of a particular class of concepts and is composed of a hierarchy of schemata embedded within schemata. The representation at the top of the hierarchy is sufficiently general to capture the essential aspects of all members of the class. (Adams & Collins, 1979, p. 3)
>
> By the term 'frame' I have in mind any system of concepts related in such a way that to understand any one of them you have to understand the whole structure in which it fits; when one of the things in such a structure is introduced into a text, or into a conversation, all of the others are automatically made available. (Fillmore, 1982, p. 111)
>
> The prototype, like the frame, refers to an expectation about the world, based on prior experience, against which new experiences are measured and interpreted. (Tannen, 1993, p. 17).

Schemata, then, are packaged units of information that represent knowledge about objects, events, situations, and sequences of actions. Not only are they organized in networks as indicated by Fillmore, but they are also organized hierarchically. The network patterning permits the establishment of syntagmatic and paradigmatic associations with related schemata or frames, while the hierarchical organization allows for the possibility of combining top-down and bottom-up processing modes in understanding. As pointed out in Chapter 1, the interaction of both processing modes is crucial for an understanding of text processing as an active procedure, rather than as a passive receptive skill based on decoding. Whereas bottom-up procedures are based on decoding, top-down procedures work on higher level conceptualizations that help us to make hypotheses about the input. The importance of a combination of both modes is demonstrated by a series of experiments with second language learners where excessive reliance on one of the two modes can lead a learner to errors in processing (see Carrel, Devine, & Eskey, 1988, Chapter 7). Thus, excessive trust on top-down processing modes can lead to vagueness and imprecision and excessive reliance on bottom-up processing can lead to deficiencies in the understanding of the more general and abstract concepts. Carrel and colleagues (p. 79) distinguish between *content* schemata or *world* schemata and *formal* schemata, which correspond roughly to the distinction that Cook (1994) establishes between *world* schemata on the one hand and *language* and *text* schemata on the other.

Schemata are viewed by Rumelhart as providing the "skeletons around which the situation is interpreted" (1980, p. 37). In this way, default elements are provided by schemata, and particular variations characterizing specific situations can be indicated within the general "skeleton" provided by the schema.

The notions of schema and frame rest on the notion of expectation because schemata and frames are used to understand new experience by making predictions and hypotheses about the new situations. These are then compared to previous experiences collected in schematic packages (see Carrell, Devine, & Eskey, 1988; Rumelhart, 1980, pp. 38–39; Tannen, 1993, p. 16). This leads to a dynamic view of schemata as discourse units, although the dynamic perspective is by no means common to all schema theories. An example of a dynamic view of the function of schemata in discourse processing is Rumelhart's theory, where schemata are considered as:

active computational devices capable of evaluating the quality of their own fit to the available data. That is, a schema should be viewed as a procedure whose function it is to determine whether, and to what degree, it accounts for the pattern of observations. (p. 39)

Rumelhart points out that top-down and bottom-up processing modes are combined to obtain "enough evidence in favour of a schema" (p. 42).

There is a close relation between the notions of schema or frame and the one of prototype. As Tannen (1993, p. 17) points out, both have to do with expectations and knowledge of the world. Thus, in the Anglo-Saxon population, a word like robin is a prototype of the family BIRD, which creates expectations in English speakers regarding what a bird should look like by means of a paradigmatic comparison to other members of the class. The establishment of connections with other members in a class is also typical of schematic relations, which are exemplified by Fillmore (1982, p. 111) by means of the word son, which, in order to be understood, requires the understanding of the whole network of social relationships present in syntagmatically associated frames, such as father, mother, sister.

Semino (1994, 1997) points out that the difference between prototype theory and schema theory lies in the objectives of analysis of the two disciplines. Whereas prototype theories (as in Rosch, 1973) are concerned with the organization of experience regarding objects, persons, or actions, schema theories are concerned with more complex relations, such as the representation of sequences of actions, for example, the RESTAURANT script. As Rumelhart (1980, p. 33) argues, a schema theory embodies a prototype theory. Similarly, Fillmore (1982, pp. 117–118) observes:

> One generalization that seemed valid was that very often the frame or background against which the meaning of a word is defined and understood is a fairly large slice of the surrounding culture, and this background understanding is best understood as a "prototype" rather than as a genuine body of assumptions about what the world is like.

On the basis of the previous observations, the goal of schema theory can be described as that of accounting for "how knowledge is represented and how that representation facilitates the *use* of the knowledge in particular ways" (Rumelhart, 1980, p. 33). Similarly, schema theory should also account for the relation between reader and text in text comprehension. Thus, Adams and Collins (1979, p. 3) consider the main function of schema theory to be the following.

> The goal of schema theory is to specify the interface between the reader and the text—to specify how the reader's knowledge interacts with and shapes the information on the page and to specify how that knowledge must be organised to support the interaction.

Cook (1994, p. 27) also stresses the significant role played by the reader in schema theoretic frameworks, and for this reason, considers them adequate complements to more formally oriented discourse theories. This issue is expanded in Section 4.3.2.

To sum up, schema theory can provide useful insights regarding the following points.

1. The way in which expectations are created and defeated or confirmed in discourse;
2. The way in which reading and understanding of a (literary) text is not a passive activity, but an active dynamic process.

Schema theory also has serious drawbacks, however, such as the lack of specificity regarding the number and types of schemata that can exist or be activated at one particular point, the overlap between different categories of schemata and between different levels in hierarchies of schemata, and, as Cook (1994, p. 74) and Emmott (1994, p. 157) point out, the tendency to overlook the relations between schemata and specific linguistic items or structures. In general terms, schema theories lack the necessary constraints for a scientific theory to be fully developed and to make it testable (see Thorndyke & Yekovich, 1980). As Thorndyke and Yekovich add, however, the theory also has enough flexibility and capacity of insight to be a useful instrument of analysis.

4.2.3. Fillmore's Frame-Semantic Approach to Negation

Little work has been carried out on the possible application of schema theoretic or frame semantic principles to the understanding of negation in discourse. The few works that deal with negation from this perspective typically deal with simpler versions of schema theories, where no specification is made of a hierarchy of schema types and their functions. These studies have significant contributions to make to the understanding of the cognitive properties of negation.

The view of negation as involving a relation between negative and positive terms, which are conceptualized by means of frames or schemata, is defended by studies such as Shanon (1981), Fillmore (1982, 1985), and Pagano (1994).[1] In this section, we discuss the following issues from a frame-semantic point of view: (1) the criteria for the appropriateness of negative utterances in a context; and (2) the characteristics of negatives in Fillmore's work: context-free and context-bound negation and within-frame and across-frame negation.

A frame-semantic approach to negation may account for the appropriateness of some utterances and the inappropriateness of similar ones in identical contexts, where no syntactic reason for their acceptability or unacceptability is involved. The authors mentioned argue that the fact that sentences like those in (1) are appropriate, while those under (2) are inappropriate, can be explained by reference to the frames that are operating in each case.

(1) a. There's no furniture in the room.
    b. The picnic was nice but nobody took any food.

(2) a. ? There are no diamonds in the room.
     b. ? The picnic was nice but nobody watered the grass.

Shanon (1981, p. 42) argues that (1)a is acceptable because furniture is part of the ROOM frame, however, (2)a is not acceptable or sounds odd because diamonds is not necessarily part of the ROOM frame. Similarly, Pagano (1994, p. 257) argues that (1)b is an acceptable utterance because food is part of the PICNIC schema, however (2)b sounds odd because watering the grass is not something that one usually associates with picnics. The view presented by these authors is that for a negative utterance to be appropriate, it has to operate within an activated schema or frame. Such frames or schemata might be activated by specific lexical items in the discourse, such as room and picnic in the examples, but they might also be shared assumptions concerning cultural behavior, beliefs, and in general shared knowledge that is not explicitly expressed in the discourse. Thus, Pagano (1994, p. 256) points out that the utterance in example (3) denies an assumption that is part of our shared knowledge about a world where brides wear white dresses at weddings.

(3) The bride was not wearing a white dress.

Utterances of this kind reveal aspects that are variable across cultures. Shanon (1981, p. 42) provides a similar example where a cultural schema is activated not by a particular item in the discourse, but from the particular situation. In this example, the situation involves the activation of a typical RESTAURANT script, where waiter is a constitutive part.

(4) A: Why did you pick your food yourself?
     B: Because I saw no waiter.

Turning now to Fillmore's (1982, 1985) view of negation, the author first distinguishes between frames that are *evoked* by the text (e.g., the ROOM and PICNIC frames) and frames that are *invoked* by the reader to make sense of the text (Fillmore, 1982, p. 124). The latter are "genre-culture specific frames independent from the text," such as, for example, the Japanese tradition of starting a letter by making a comment on the current season. According to Fillmore (1982, p. 124), the reader invokes a schema for letter-writing in Japanese, which he or she applies when reading a Japanese letter, and this enables him or her to make sense of the reference to the season. What is not clear is if these invoked schemata are exclusively genre-related or if they are actually independent from the text; it can be argued that specific lexical items in the text also *evoke* the current season schema in the reading process. The question is, rather, whether this schema, CURRENT SEASON, is part of the higher level schema LETTER or not, as is argued by Shanon (1981) and Pagano (1994) in the previous examples.

Fillmore (1985, pp. 242–245) further discusses the properties of negation from a frame-semantic approach by considering (1) the differences between what he calls *context-free negation* and *context-dependent negation*; and (2) the differences between *within-frame negation* and *across-frame negation*. As an illustration of the difference between *context-free* and *context-sensitive* negation, Fillmore provides the following examples.

(5) a. Her father doesn't have any teeth.
    b. Her father doesn't have any walnuts.

According to Fillmore's (1985) notion of context, which seems to coincide with that of co-text or immediately preceding discourse, (5)a is context free, in that the expression of the negative does not need to be cohesively related to a previous item in the discourse because the frame for a person's face is always available to us. In (5)b, however, negation has to be interpreted in relation to some previously uttered discourse of which the frame WALNUTS is a part. The distinction between context-bound and context-free negation in these terms is not sufficiently clear. It seems that both examples are context-bound types, although the context dependency operates at different places in discourse in each case. In (5)a the dependency is on a word within the same sentence. In (5)b it is on a word supposedly outside the boundaries of that sentence. This can be explained because, in fact, the acceptability of (5)a versus the oddity of (5)b in absolute terms still obeys the same principles as those argued for in examples (1), (2), and (3). Both sentences in (5) have the word father, which evokes a particular frame. Example (5)a is easily understandable because teeth forms part of the frame FATHER, as FATHER contains the feature human, which contains the attribute has teeth. However, (5)b is not understandable outside a broader context because WALNUTS is not a property of FATHER in the way TEETH is. Thus, utterance (5)a denies a part of a frame that is present in the utterance itself; however, (5)b denies an item that is not present in the utterance and consequently, has to be recovered from previous discourse.

With regard to the distinction between *within-frame* and *across-frame* negation, it reflects the same kind of dichotomy that is presented between predicate and metalinguistic negation when dealing with syntactic negation (see Section 2.2.1. in Chapter 2). The difference is illustrated by Fillmore (1985, p. 243) by means of the following examples.

(6) a. John isn't stingy. He's generous.
    b. John isn't stingy. He's downright thrifty.

Whereas (6)a is an example of within-frame negation, in that the frame STINGY is introduced and kept by establishing an opposition between the positive and the negative terms stingy–generous, (6)b is an example of across-frame negation because the frame itself is denied to introduce a different one. That is, instead of operating on a scale where *stingy* and *generous* are the polar opposites, a new frame is introduced, where *stingy* and *thrifty* are established as opposites. The phenomenon of across-frame negation is particularly interesting if seen from the perspective of linguistic creativity. Leinfeller (1994, pp. 81–82) points out that ad hoc relations may be established in literature for stylistic effects. Leinfeller (1994) provides the following example from everyday conversation.

(7)   This is not gray. It is dirty.

In this case, we are asked to discard the set in which *gray* would be normally understood, that is, as contrasting with other colors, which would yield a sentence like (8).

(8)   This is not gray, it is white.

By saying *it is dirty*, the speaker is introducing a different set altogether, which creates a new and unexpected contrast between two apparently unrelated terms, *gray* and *dirty*. Here, again, knowledge of the world and cultural knowledge are crucial in order to understand the opposition. In example (8), it is obviously the fact that we all know that light colored surfaces go gray or brown, or in general darker, if one does not bother to clean them periodically. This phenomenon is particularly interesting when apparently incongruous oppositions are created with a humorous effect, an aspect that is discussed in the examples at the end of this chapter.

The frameworks discussed in this section provide interesting insights into semantic and contextual principles governing the use of negation, as they provide the necessary tools to tackle the question of how stored knowledge intervenes in the process of understanding the negative term, its relation to the corresponding affirmative, and its adequacy in a discourse context. The explanations, however, are limited for several reasons. First, the examples provided in this section are limited to sentences or brief exchanges; this seriously limits the possibility of creating networks of schemata, a standard process when reading a text. A related problem is that no specific hierarchy or distinction of categories of schemata is made in such a way that we can account for more complex and even conflictive examples, such as those found in *Catch-22*. For these reasons, two works in schema theory that are more elaborate and that provide powerful tools for the analysis of how schematic knowledge is organized and processed in discourse are introduced.

4.2.4.   Schank and Abelson's Scripts, Plans, Goals, and Understanding

Schank and Abelson's (1977) book is an extremely influential work that provides a comprehensive and powerful framework of schema theory. Its main objective is to develop a model of human knowledge that will also apply to artificial intelligence. Schank and Abelson take as a point of departure the notion of episodic memory (p. 17), which they define as: "an episodic memory is organized around propositions linked together by their occurrence in the same event or time span." From this, they develop the notion of script, a notion that is based on the observation that certain situations are coded as more or less fixed sequences of actions. Below is the famous example of the RESTAURANT script situation.

(9)   John went to a restaurant. He ordered a coq au vin.
      He asked the waiter for the cheque and left. (p. 39)

The word *restaurant* activates the RESTAURANT script, which contains a number of props (for example, *table, chair*), the roles of participants (*the customer, the waiter, and the cook*), the entry conditions (being hungry), results (hunger is satisfied), and scenes (entering the restaurant, reading the menu, ordering the meal, eating the meal, paying, and leaving). These elements are called *headers* (Schank & Abelson, 1977, pp. 48–49). Scripts can be of three types (p. 41).

1.   Situational scripts: restaurant, bus, jail.
2.   Personal scripts: being a flatterer, being a lover, being a friend.
3.   Instrumental scripts: starting the car, lighting a cigarette.

The main function of scripts is to provide the means of recovering the presence of default elements in the discourse when these are not expressed explicitly. Thus, because we activate the restaurant script when reading (9), we understand who is the waiter in that situation and what is his role, and we are able to infer that if John ordered coq au vin he most probably ate it, and that if he asked for the cheque, he paid for it too. At the same time, scripts also allow us to recognize variations in the default elements by means of what Schank and Abelson call *tracks*. In example (9), other tracks may be coffee shop, fast food restaurant, Chinese restaurant, and so forth.

A script is defined as "a predetermined, stereotyped sequence of actions that defines a well-known situation" (Schank & Abelson, 1977, p. 41). The definition reveals that, being stereotyped entities, they are useful in handling everyday situations but they cannot deal with unfamiliar or totally new situations. The distinction is not clear-cut, but rather a cline. An experience may be new once, but after several occurrences, it may be eventually stored in the form of a script.

It is important to realize, however, that not all connected pieces of text reveal script-like structures. The following is an example of a connected text that is not a script.

(10) John wanted a newspaper. He found one in the street.
      He read it. (Schank & Abelson, 1977, p. 39)

Continuing now with the characteristics of scripts, another significant feature is that a script must be written from the point of view of a particular role. That is, the RESTAURANT script, for example, must be written from the point of view of the customer, the waiter, the cook, or some other person. Some scripts are not fully activated during comprehension, but may be partly instantiated only. This is the case of *fleeting scripts* (Schank & Abelson, 1977, p. 46). In the previously discussed restaurant script, fleeting scripts can be evoked by headers in the script. In order for a script to be nonfleeting, at least two headers or two lines from the script sequence must be activated (p. 46).

In addition to scripts, Schank and Abelson (1977) consider that there are other higher level structures that intervene in the process of understanding situations that are not stereotyped. Thus, plans are used in situations for which there is no available script: "A plan is made up of general information about how actors achieve goals" (p. 70). Whereas scripts are specific, plans are more general and they enable us to identify goals. The authors provide the following example.

(11) John was lost. He pulled up his car to a farmer who was standing by the road. (p. 75)

The notion of plan allows for an identification of a purpose in John's stopping to ask the farmer, by means of inferencing a goal (know). By means of a planbox (ask), we understand John's plan to ask the farmer for directions in order to get to know his way. If we are not able to identify plans and goals, it may be difficult to make sense of a text, as in (12).

(12) John was lost. He noticed a chicken. He tried to catch it. (Schank & Abelson, 1977, p. 76)

Here, it is extremely difficult to identify the reasons for John's behavior, as there seems to be no connection between the fact that he is lost and the fact that he notices a chicken and tries to catch it, unless he has been lost for a very long time and he is starving, for example. In any case, the difficulty in understanding the text as a whole is linked to the lack of a unifying goal and plan.

Goals, then, are also basic to understanding, and they constitute the level above plans on Schank and Abelson's hierarchy. Goals can be of different types (Schank & Abelson, 1977, pp. 113–119).[2]

1. S–satisfaction goals: of hunger, sleep, sex.
2. E–enjoyment: travel, exercise, sex.
3. A–achievement: possessions, power, job, skills, social relations.
4. P–preservation: health, safety, position, property.
5. C–crisis: (a special class of P-goals): health, fire, storm.

6. I–instrumental goals: goals that are instruments in order to achieve other goals.
7. D–delta goal: similar to I-goals, only that they involve scripts.

Goals come from themes, which are defined as follows: "A Theme is essentially a generator of goals. When a theme is identified, it makes sense of a person's behavior by providing a prior context for his actions" (Schank & Abelson, 1977, p. 119). There are three types of themes: role themes, interpersonal themes, and life themes. Themes make it possible to identify actors' goals and to make predictions about future goals. Some role themes are institutionalized (waiter, President), others are not (customer, messenger). Examples of interpersonal themes are friend, lover, enemy. Life themes describe the general aim in a person's life, like being rich or honest. The authors provide an example of the luxury living life theme and its associated goals (pp. 147–48):

LUXURY LIVING LIFE Theme
- Theme recognizer patterns: e.g., stay at smart hotels.
- General goals: e.g., have desirable objects.
- Instrumental goals: e.g., work hard.
- Production rules: e.g., if there's an opportunity for money, take it.

The four types of schemata developed by Schank and Abelson are organized in the hierarchy

- Themes
- Goals
- Plans
- Scripts

The advantages of a framework of this kind against simpler versions of schema theory is summarized in the following points.

1. The establishment of a hierarchy of schemata where failure to understand at one level may be solved at a higher level of understanding.
2. The identification of stereotyped sequences of actions, which the authors call scripts, make it possible to recover the implicit presence of default elements not explicitly mentioned, and to present recognizable variations of these default elements.

The framework also presents disadvantages, however, which are found in the following aspects.

1. The rigid character of the categorizations, as they do not account for schema change; and

2.   The overlap between the categories.

Also, as Semino (1994, p. 138) points out, some of the observations, such as the types and characteristics of goals, are typical of a very specific identity, that is, a white, male, middle-class American, so that they may not be valid for other social groups. These disadvantages do not prevent the framework from providing useful tools for the analysis of how knowledge is stored and processed in communication; however, Schank and Abelson's model will be complemented with Schank's (1982) *Dynamic Memory*, which focuses on the dynamic aspects of discourse processing.

4.2.5.   Schank's Dynamic Memory

The dynamic aspects of schema production and change, especially in learning about new situations, are the focus of Schank's (1982) work.[3] Schank takes the notion of *reminding* as a point of departure, and observes the way in which many events in our experience remind us of others, sometimes by establishing connections between apparently unconnected areas of experience. Thus, one kind of restaurant reminds us of another, even when there are changes in the sequences of actions. Schank (p. 23) provides the example of a restaurant where one is asked to pay before eating. In order to account for knowledge in these terms, Schank and Abelson's (1977) model proved insufficient, as the categories established were not flexible or general enough to account for changes in scripts such as the one mentioned previously, or for how unrelated schemata may be connected through reminding. Schank provides an example situation where a man had been standing for a long time in a queue to buy only one stamp reminded him of the people who stop at petrol stations to buy only a few liters of petrol (Schank, 1982, p. 32). According to Schank, reminding is goal based, which means that we are reminded of other scenes by means of connected goals, in the case of the example, the lack of fit between the goal of purchasing something and the goal of being maximally efficient.

Reminding is very much failure driven, and a view of understanding and learning must account for this point. Schank (1982, p. 46) argues for a view of memory and understanding where we have expectations about certain events and make predictions about them. When the expectations and predictions are defeated, however, we "write down" the error and we remember. These variations from expectations are remembered when parts of the relevant scene are activated in other contexts. This account requires a dynamic notion of script, which can change in response to new input (p. 82).

Schank (1982, p. 15) makes an important distinction between two main types of schemata (which constitutes the basis of the flexibility of his system), structures and organizers of structures. The idea is that lower level schemata do not form part of fixed sequences, such as scripts in the sense used in the (1977) work, but they can be activated by different higher level units depending on the situation. These higher level units are organizers of structures. Thus, the RENT-A-CAR scene will be activated by means of higher structure, for example TRIP, "as in itself it does not contain the reasons for itself" (1982); furthermore, it will not belong to only one fixed sequence, as was the case with the RESTAURANT script in Schank and Abelson (1977). A clearer example may be the HOTEL ROOM scene, which may belong to many different higher level structures. To this respect, the author points out that "episodes are not remembered as wholes but as pieces" (Schank, 1982, p. 90).

There are two types of structures: scenes and scripts. Scenes are general in character and scripts are specific. The difference from the notions in the (Schank, 1977) model is that neither scenes nor scripts "exist in memory as a precompiled chunk" (Schank, 1982, p. 16). Rather, the different parts may be reconstructed depending on the situation one is in. In this sense, if we talk, for example, about a visit to the dentist, we do not have a VISIT TO THE DENTIST fixed script,[4] but information of two kinds: information about what other scenes or general structures comprise a visit to the dentist, and specific information, or colorations, to each scene, which indicates the differences between aspects such as a dentist's waiting room versus a lawyer's waiting room.

There are two kinds of high level structures: memory organization packets (MOPs) and thematic organization packets (TOPs). High level structures, in general, are formed by making generalizations about related areas of experience. Thus, from a visit to a dentist we can make generalizations about doctors, and from doctors we can make generalizations about health care service. HEALTH CARE SERVICE can be an MOP. This unit is defined by Schank as "Information about how memory structures are ordinarily linked in frequently occurring combinations, is held in a memory organization packet (MOP)" (1982, p. 83). MOPs are both storing and processing structures that allow us to provide place for new inputs and to provide expectations from which we can make predictions about future, related events. The differences between scenes and scripts on the one hand, and MOPs on the other, have to do with the degree of generality that characterizes the latter. Scenes and scripts are bound by the setting where they are activated, and by their stereotyped sequence of actions. A MOP covers different settings and has a purpose that is not directly obtainable from the different scenes that comprise it.

MOPs typically come in threes: personal, societal, and physical. For example, the dentist visit will cover: M-health protection, M-professional office visit, M-contract, respectively. The sharing of structures may lead to memory confusions, such as not being able to remember whether something happened at the dentist's waiting room or at some other waiting room. This is counterbalanced, however, by the advantages of MOPs in processing different types of scenes. MOPs contain the following information, which they organize (Schank, 1982, p. 90):

- A prototype;
- A set of expectations organized in terms of the prototype;

- A set of memories organized in terms of the previously failed expectations of the prototype; and
- A characteristic goal.

MOPs may contain different kinds of scenes: physical (waiting room, airport lounge), societal (contract), or personal (referring to private plans). Schank proposes a higher level structure, which is used in organizing or planning MOPs, it is called a meta-MOP. From this perspective, we can have a meta-MOP trip, covering the plans to carry out a series of goals (get resources, make arrangements, etc.). This meta-MOP is used to construct MOPs such as airplane, which is related to other subgoals, such as, book ticket, check in, and so forth.

If we summarize what we have up to now, we can establish this hierarchy:

Meta-MOPs

MOPs

Scenes

Scripts

The boundaries between the different levels may not always be clear, but this does not invalidate the operational value of the system. Thematic organization packets are high level structures that store information that is independent from particular domains. In this sense, they are abstractions from actual events, which enables us to establish connections between different events and find similarities between them. This phenomenon lies behind our capacity to be creative in our understanding. Schank (1982, pp. 111–112) provides the example of the word *imperialism*, which is used about countries in international relations. The term can be used to describe someone's attitude to land possession or other possessions, however, and we will easily understand what is being referred to. Schank provides several examples of how TOPs are organized. An example is shown in Table 4.1., where *West Side Story* reminds us of *Romeo and Juliet* (p. 113).

This pattern allows for the identification of some of the main elements in TOPs, in particular the goal types and their related problems. This view of understanding can be particularly interesting when applied to the understanding of literary texts, where reminding processes of this kind, which involve the association of apparently disparate domains, take place very frequently.[5] This is also true of humorous situations, which often arise precisely by means of establishing an unexpected

TABLE 4.1.

Reminding	Goal	Conditions	Features
West Side Story/ Romeo and Juliet	Mutual goal pursuit	Outside opposition	Young lovers False report of death

connection between two different domains. The humorous potential of certain associations is mentioned by Schank (1982) with regard to some of the examples provided.

4.3. SCHEMA THEORY AND LITERATURE

Approaches to literariness based on schema theoretic principles have developed by means of the influence of work in artificial intelligence and cognitive psychology. Findings in these fields have created an interest in literary works as text types that tend to challenge existing schemata, and can lead to schema change and the creation of new schemata.

The following sections focus on a dynamic perspective of schema activation and use in literary discourse, in particular, on a discussion of Cook's (1994) contribution to this area of research, but a brief overview of work that has been carried out in this line is given.

4.3.1. Schema Theories of Literariness

Schema theories that deal with literature have been mainly concerned with the function of schemata in the elaboration and processing of stories, called *story grammars,* from a schema theoretic perspective (see the special issue on this topic in *Text*, van Dijk, 1982). Although these works have been very significant in the development of artificial intelligence and its application to the understanding of text coherence, the applications to the nature of literariness are limited for several reasons. The texts used are typically simplified versions of popular stories, which present easily recognizable structures and variations that can be handled by a computer. This means that complex literary works have not been analyzed in this way, mainly because they could not be dealt with by a computer program. Cook (1994) observes these deficiencies in schema theory as applied to literature and argues that it is necessary to develop a theory that may account for the way in which new schemata, or radical departures from already known schemata, interact with old schemata to yield complex literary works.

The way in which this process may take place is outlined by de Beaugrande (1987), who observes that "the most famous 'literary' stories—those that survive, like *The Arabian Nights* and *The Decameron*—are those which offer stimulating mixtures of confirmation and violation of what people expect." These expectations can be said to be organized in schematic units. More precisely, de Beaugrande points out (p. 56) that literature is a type of "communicative domain in which certain top-level schemas . . . control the selection, activation or formation of lower-level ones." In this view, the highest level schema in literature is what de Beaugrande calls ALTERNATIVITY, which he describes as a state where "the participants in the literary communication are free and willing to contemplate other worlds beside the accepted 'real world.'" Interestingly, de Beaugrande's view brings together notions from text world theory (in his proposal of a theory where a literary work is said to present an alternative world) and schema theory (in the stress given to the role of the reader in the process of understanding the literary work by means of schema processing). For de Beaugrande (p. 60), fictionality is a relative notion that is established by comparison with actuality. There may be a great deal of variation regarding the degree of overlap between a fictional world and the real world, as between historical narratives and fantastic literature. According to de Beaugrande (p. 60), however, all literary works "are concerned with the pre-conditions of reality," in the sense that, in order to understand both imitation of principles of reality and their violations, we need some kind of reference to the characteristics of reality, to what is familiar to us. De Beaugrande's approach is attractive because it proposes a way of combining the notion of alternativity typical of text world theory with the hierarchy of schemata in text processing and understanding that is typical of schema theory. Semino (1994, 1995, 1997) develops a model that precisely combines these two principles and applies them to the interpretation of poetry, a literary genre that tends to be overlooked by both schema theories and text world theories, which tend to focus on fiction.

Other attempts to incorporate schema theoretic principles in an understanding of literary discourse are Miall and Kuiken (1994, 1998), who combine the notions of defamiliarization and the role of the reader in text processing typical of schema theories, and Müske (1990), who defines literariness in terms of the notions of frame and superstructure.

Emmott (1994) uses the term *frame* to stand for particular mental constructs used in the understanding of characters and locations in fictional works. In her framework, the frame is not a package of information, but is understood as "a tracking system which monitors which particular characters are 'present' in the location at any one point." (p. 158). This argument is based on the observation that there are certain features of narrative discourse that need an explanation based on cognitive principles (p. 157), such as, for example, how reference is assigned by readers in cases where the relation with an antecedent is not expressed explicitly. More interestingly for the present discussion, Emmott also points out that cognitive modeling is basic in the construction of a fictional world and the processing of flashback (p. 157). The author provides an example where a flashback is introduced by means of the past perfect, although there is an immediate switch to simple past. Emmott (p. 161) explains this as "The reader knows, however, that these sentences denote flashback events because on entry to the flashback s/he has set up a flashback frame." Further, Emmott (p. 157) also establishes a distinction between what she calls *general knowledge mental structures* and *text-specific mental structures* (compare Cook's [1994] *world schemata* and *language* and *text schemata*). According to Emmott (1994) more attention has been paid to the explanation of general knowledge mental structures. Text-specific ones have been little explored, although they should be of great interest to discourse analysis. Emmott proposes an analysis based on the notion of frame defined as a means of accounting for certain text-specific phenomena, such as reference and flashback.

4.3.2. Cook's Model and the Function of Cognitive Change in Literature

Cook's (1994) framework is an attempt to bring together principles from stylistics in the formalist tradition and schema theory. From the former, Cook adopts the notion of *defamiliarization*, which he adapts to a discourse theory, which also accounts for the role of the reader.[6] This aspect is the most significant one adopted from schema theories in general. According to Cook (p. 65), some schemata are not mental representations built on different codes from language but are located in the language itself. From this perspective, defamiliarization can apply at least at three levels of text understanding, which correspond to the three levels in the hierarchy of schemata in Cook's framework (p. 181): *language schemata, text schemata*, and *world schemata*. As explained previously, language and text schemata correspond roughly to what in the literature is typically referred to as *content* schemata. World schemata, however, receive the same name, or, otherwise, are referred to as *formal* schemata. The main concern in Cook's model is to show how deviance at the levels of language and text can lead to deviance in world schemata, and cause restructuring and change of schemata.

The notion of deviance used by Cook (1994) is borrowed from the formalists' notions of deviance, together with the notions of foregrounding and defamiliarization. It is adapted, however, to account for schema change in the reader as well. Cook (p. 182) points out that certain types of discourse, typically literary texts, can be said to have the specific function of inducing schema challenge and, possibly, schema change in the reader. This is possible in literary texts because it is a discourse type that is not directly concerned with what are standardly understood to be more practical communicative uses in society. Literature is not bound by the need to be communicatively efficient. This is what enables it to be more challenging to established conventions. In this sense, literary texts can be said to display a function that cannot be included in either the ideational or interpersonal functions of language in Halliday's (1994) model, or identified with Jakobson's (1964) poetic function. Cook puts forward a view where the main function of literary texts (though the function is by no means exclusive of this text type) is that of *cognitive change*. This function is described by the author as follows.

> some discourse is best interpreted as though it followed a maxim 'change the receiver'—though that may not necessarily have been the intention of the sender. Such discourse fulfills the need to rearrange mental representations: a process which can be best effected in the absence of pressing practical and social constraints. . . . In some discourses, in other words, language has a function not accounted for in the

functional theories referred to above: the function of changing mental representations. (p. 44)

Although the idea that literature provides a new means of interpreting experience and idiosyncratic insights into human life and thought is certainly not new, Cook's contribution lies in the attempt to systematize how previous experience is reorganized by means of the influence of the schemata evoked by reading a text. This constitutes a significant contribution to previous discourse theories, which have been concerned with the recovery of violations of text-structural principles by means of conversational implicature (see Pratt, 1977). Furthermore, the fact that the theory relies heavily on the notion of deviance, which has received a great deal of criticism in recent years (see, for example, Carter & Nash, 1990; Fish, 1980), does not invalidate the potential of the framework to account for the way in which particular texts strike us as unusual, challenging, and difficult in different ways. The author is well aware that the notion of deviation is a relative one, as was discussed in Section 1.2. in Chapter 1. The possible weakness of the theory's reliance on the notion of deviation is compensated by the intuitively felt necessity to incorporate the notions of schema refreshment and cognitive change as basic categories in a theory of literariness.

4.3.2.1.   The notion of schema refreshment

According to Cook (1994, p. 191), from the point of view of the effect on the schematic knowledge of a reader, discourse can be classified into three main types: schema reinforcing, schema preserving, and schema disrupting (this last one leading to schema refreshment). Schema reinforcing and preserving discourse is discourse that confirms and reinforces already existing schemata, as in the examples of advertisements discussed by Cook. In these cases, even if there might be deviance at the levels of language use and text structure, the schemata that are evoked concerning behavior in social relationships, including buying products, are extremely conventional.

Schema disrupting discourse, on the other hand, does not reinforce preexisting schemata, but either destroys old schemata, constructs new ones, or establishes new connections between already existing schemata. Cook argues that these processes are the basis for schema refreshment. Cook (1994, pp. 192–193) then describes the characteristics of schema refreshment, which is presented as a relative concept subject to reader variation and change throughout time. This accounts for the fact that schema disrupting texts at one point in time may be incorporated in the canon later and become schema preserving. The author gives the example of Jane Austen's novels, which in her time were greatly innovative, but are now examples of conventional novels.

An important feature of schema refreshment is that it takes place in the interaction between levels and not at one particular level in isolation.[7] Cook (1994, pp.

197–198) describes schema refreshment as a discourse phenomenon that is the basis of his notion of discourse deviation.

> Where there is deviation at one or both of the linguistic and text-structural levels, and the deviation interacts with a reader's existing schemata to cause schema refreshment, there exists the phenomenon which I term "discourse deviation." (p. 198)

Consequently, the task of a theory of discourse deviation such as Cook's is to make explicit the connections between the deviations at the text and language levels on the one hand, and the changes in the schematic representations of the world in the reader on the other. The process of discourse deviation is a dynamic one where the reader maps his or her representations of language, text, and world schemata against the corresponding ones evoked by the text. It is an ongoing process that involves a constant up-dating and restructuring of information (compare Werth's version of discourse processing in Chapter 3). To illustrate the framework, Cook analyzes a series of texts and classifies them according to the types of deviance found in each of them. For example, William Blake's poem *The Tyger* is classified as containing deviations at the lexicogrammatical level, but with a conventional text structure corresponding to a ballad. The deviations at the language level induce a restructuring of world schemata that have to do with the nature of God and evil, among other things. *The Turn of the Screw*, on the other hand, is described as having conventional language schemata and deviant textual schemata related to the presence of an unreliable narrator. This leads to schema refreshment by means of creating a deliberate ambiguity, which makes the reader question previous assumptions about the reliability of narrators and the major themes of the work and how they are treated. A problematic example discussed by Cook is Bond's poem *First World War Poets*. Cook (1994, p. 201) argues that in this case, discourse deviation is produced by means of a combination of *ordinary language* and a conventional poetic layout: "It is the absence of text-structural and linguistic deviation which, combined with the expectations set up by poetic form, 'represents' the schema-refreshment advocated by the poem." (p. 201).

This view overlooks the fact that there is something striking about the language of the poem, not only the schemata evoked by it. It appears that Cook's framework does not pay enough attention to the significance of the hierarchical nature of schemata and the dependency of the lower levels on the higher levels. By this is meant that the higher levels should be understood to determine the lower levels, so that the schema for a conventional poetic structure determines the following of certain conventions regarding the lexicogrammatical patterning of the poem. According to this, Bond's poem would be deviant at the lower level of the lexicogrammatical schemata, precisely because the language goes against the expectations a reader has regarding what should be "poetic language" and its content. This leads to a view of discourse deviation as hierarchically organized and truly reflecting the relations between the different levels.

Finally, it has to be pointed out that more research needs to be carried out in order to specify more precisely how schema refreshment takes place. To this respect, Semino (1995, p. 104) observes that "schema change is not only infrequent, but also hard to verify," and suggests a partial redefinition of the notion of schema refreshment so that it can be applied to less dramatically challenging texts than those analyzed by Cook. Furthermore, other factors need to be explored, such as the motivation of different individuals according to criteria such as age, gender, race, cultural background, and education, as they might prove to be crucial in the determination of schema change in an individual.

4.4. A SCHEMA THEORETIC INTERPRETATION OF HUMOR

If we consider the description of negation given in Chapter 2, where I argue for a cognitive approach to the phenomenon, in such a way that the understanding of a negative term involves the defeat of an expectation, we can see there is a close connection with the phenomenon of humor, which also works as the defeat of an expectation (see Freud, 1976b, Norrick, 1986; Shulz, 1976).[8] Many of the examples of extracts illustrating the function of negation in *Catch-22*, especially those involving contradiction, have a humorous effect—at least for some readers like the author of this book; for this reason, a discussion of the characteristics of humor as incongruity could shed light on the process of understanding certain uses of negation. In this section, we discuss some crucial notions related to the view of humor as incongruity and the adequacy of schema theory to account for the phenomenon.

In the analysis of humorous effect, we are particularly interested in two aspects, namely (1) the fact that the humorous effect arises as the defeat of an expectation; and (2) the fact that humor takes place when the incongruity is perceived to have a further meaning. For this purpose, we concentrate on the view of humor as incongruity (see Freud, 1976b, 1966; Norrick, 1986; Shulz, 1976; Simpson, 1989). Incongruity in jokes is defined by Shulz (1976, p. 12) as "a conflict between what is expected and what actually occurs in a joke." Shulz provides several examples of how the humorous effect takes place in different text types by means of ambiguity in the lexicon, in phonetic or structural aspects, as in Groucho Marx's saying in (13).

(13) I ought to join a club, and beat you over the head with it.

In (13), incongruity and the humorous effect hinges upon the ambiguity of the word *club*, and the awareness that each of the two meanings is being projected in different parts of the sentence. Similarly, the second part of the sentence can be said to defeat expectations created by uttering the first part, since in the second part of the sentence a different meaning of *club* is introduced, thus producing a dramatic change in what the utterance is actually about.

Many theories of humor (see Apter, 1982; Attardo, 1994, Chapter 4; Chapman & Foot, 1976; Freud, 1976; Norrick, 1986) defend a view that consists in the acceptance of the incongruity and its resolution at a higher level of processing. Indeed, Shulz argues (1976, p. 13) that the higher level resolution of incongruity in humor is what differentiates it from nonsense, where the conflict remains unresolved. The author observes that "whereas nonsense can be characterized as pure or unresolvable incongruity, humor can be characterized as resolvable or meaningful incongruity." (p. 13). The process is defined as one where the subject is first aware of the incongruity and subsequently, searches for a resolution of the incongruity.

This view is also defended by Norrick (1986) in an analysis of humor, where he applies the notion of *bisociation* borrowed from Koestler, and combines it with schema-theoretical principles to account for how humor is produced. According to Norrick (p. 226), humor involves the phenomenon of bisociation, which he describes, quoting Koestler (1964) as follows.

> the perceiving of a situation or idea L, in two self-consistent but habitually incompatible frames of reference M1 and M2. The event L, in which the two intersect, is made to vibrate simultaneously on two different wavelengths, as it were. While this unusual situation lasts, L is not merely linked to one associative context, but bisociated with two. (p. 35ff).

According to Norrick, this phenomenon can be captured adequately by frame semantics, as each of the two frames of reference can be seen as conceptualizations which contain schematic knowledge (p. 229). Norrick considers schema theory particularly adequate to explain humor because the schema conflict that creates incongruity at a lower level can be interpreted as being meaningful at a higher level of processing, an approach to understanding that is based on the notion of a hierarchy of schemata. Norrick (p. 230) further specifies:

> This leads to a hypothesis associating funniness with schema congruence revealed at higher level. The idea of higher-level schema fit, in combination with lower-level schema conflict, lends substance to the traditional definition of humor as "sense in nonsense" or "method in madness."

The idea that the conflict between two opposite parts gives rise to incongruity and humor is also present in the theory of cognitive synergies discussed in Chapter 2. In this theory, humor can also be defined as a synergy, in that it constitutes a phenomenon where two incompatible terms are combined in such a way as to produce an unexpected effect, which is different from the two terms in isolation. The unexpected effect can be termed the humorous effect. Humor may take place in two different ways, either through a transition from a state A to a state B, as in the previous example from Groucho Marx, or in a context where the subject is aware of the two conflicting meanings from the beginning. The latter form of humor is typical of make-believe humorous situations, as when a male comedian pretends to be a woman and exaggerates his supposedly feminine attributes.

To sum up, humor can be described as a phenomenon that involves the defeat of an expectation in such a way that a conflict arises between two opposing complex schemas. The conflict is manifested at a lower level of processing, but may be resolved at a higher level, where it yields the humorous effect and is schema refreshing. The understanding of the resolution is not only variable across individuals but it is also culture dependent. This provides strong support for the argument that the higher level resolution requires the activation of specific schemata related to knowledge of the world. This explains why some readers find nonsensical what others find funny. In this respect, Freud (1976b, p. 162) points out that not all human beings are capable of developing a sense of humor, which is a precious and rare gift.

4.5. AN APPLICATION OF SCHEMA THEORETICAL APPROACHES TO THE ANALYSIS OF EXTRACTS FROM *CATCH-22*

As was pointed out in Chapter 2, in *Catch-22*, negation often occurs in association with a humorous effect. It can be argued that humor is a form of schema refreshment, and that its recurrent presence in *Catch-22* contributes to the challenging of a reader's schemata during the process of comprehension. The discussion that follows in this last section is meant to illustrate the relation between negation and humor, and how schema theoretic models such as the ones discussed in the present chapter can be used to account for such a relation.[9] The determination of the humorous effect is, of course, totally subjective, so that I rely on my own reactions to utterances as funny or not. The types of negatives analyzed in this section include both forms of what I define as syntactic negation and lexical negation (see the relevant classifications in Chapter 2).

Below is an example from *Catch-22* that presents the opposition between the contrary terms *sane* and *crazy*, which in terms of polarity can be distinguished as the positive and the negative term, respectively.

> (14) "Do you really want some more codeine?" Dr. Stubbs asked.
> "It's for my friend Yossarian. He's sure he's going to be killed."
> "Yossarian? Who the hell is Yossarian? What the hell kind of a name is Yossarian, anyway? Isn't he the one who got drunk and started that fight with Colonel Korn at the officers' club the other night?"
> "That's right. He's Assyrian."
> "That crazy bastard."

"He's not so crazy," Dunbar said. "He swears he's not going to fly to Bologna." "That's just what I mean," Dr. Stubbs answered." That crazy bastard may be the only sane one left." (p. 144)

In this extract, two opposite properties (*sane–crazy*) are attributed to the same entity (Yossarian). If we consider the opposition at a low processing level, that is, just as a contradictory attribution of two properties, the extract will be nonsensical. What I argue in the following discussion is that language uses of this kind require the interaction between this low level awareness of a contradiction and a higher level point of resolution, where the contradiction is understood as having a meaning.

If we apply Fillmore's analysis to negation in terms of frame semantics, we may observe that crazy and sane correspond to two frames that contrast as a form of within-frame negation. This view, however, is limited because it does not explain why the two terms co-occur in discourse. To account for this, we need a model that will account for hierarchies of schemata. In general terms, the higher level meaning is recovered by means of establishing the domains where the attributes *sane* and *crazy* are applicable: Yossarian is crazy because he has dared to start a fight with one of the higher officers and he is sane because he refuses to fly to Bologna, where he fears he will be killed.

In the terms defined by Schank and Abelson's (1977) model, this can be interpreted by referring to different role themes and the corresponding goals expected in each role. From this perspective, Yossarian is crazy in his role as a soldier because he goes against military orders and behavior (a soldier must be respectful towards higher officers; a soldier must always be willing to go into combat; his main goal is an S-satisfaction goal, to defeat the enemy and win the war, by means of an I-instrumental goal, fight against the enemy). As a human being, however, he is sane, because his main goal is a P-preservation goal, to survive.[10] Schank and Abelson's model provides some examples of how complementary and even apparently conflicting roles can take place in the same situation, but the model does not provide a higher level structure that makes it possible to understand in more general terms why those roles and their corresponding goals are conflictive.

At this point, Schank's notion of MOP as a higher order structure that organizes lower level goals, scenes, and roles, may be more useful. It can be argued that in this extract, as in many others in *Catch-22*, a meta-MOP WAR is operative, with a series of connected MOPs that organize information regarding societal, personal, and physical aspects about war. This is shown in Table 4.2 as follows.

As is shown in Table 4.2, in extract (13), we face at least three relevant MOPs: a physical MOP, M-AIR MISSION; a societal MOP, M-BOMBARDIER; and a personal MOP, M-SURVIVAL; each with its prototypical goals. Again, we have conflictive goals manifested in the opposition between the societal and the personal aspects. It is important to realize, however, that these roles are brought together under the higher level structure that collects information about war in general terms,

TABLE 4.2.
meta-MOP: War MOPs

MOPS	M-Air Mission	M-Bombardier	M-Survival
Goals	Bomb city	S-Bomb town S-Win war	P-Preserve life
Evaluation of goal	—	Negative: crazy	Positive: sane

and allows us to infer that although a soldier must be willing to die for his country, as a human being, he may have doubts, an aspect that is foregrounded in Yossarian's characterization. Thus, we can say that there is a priority given to the goal P-PRESERVATION OF LIFE, although the extract exploits the awareness that this priority obviously goes against the priorities set up by the military system. The assignment of the terms *crazy* and *sane* is based precisely on the goal that is given priority, with the consequent evaluation of the goal depending on the priority chosen. This might suggest, as Semino (1994) points out, that a further category collecting affective and evaluative aspects may be necessary.

From the point of view of Cook's (1994) model, what I describe as deviance at the lower level of language schemata produced by the contradictory presence of two opposing terms, produces discourse deviation and a challenge to assumptions in a reader regarding the assignment of contradictory properties to the same entity. The effect is schema disrupting, and it is schema refreshing because as readers, we become aware that one individual can be both crazy and sane at the same time, even if this would seem to be logically impossible. The acceptance of the simultaneous operation of the two properties leads to a reflection on the reasons for the development of such a conflict, which may induce a critical view of a given situation. Under normal circumstances, it should be possible to assign the properties *crazy–sane* to an individual in such a way that one will exclude the other. The fact that both properties co-exist in a given situation (war) may indicate that there is something anomalous and, consequently, not desirable about that situation. There is something wrong about war and the roles imposed by means of military structures if these roles go against the basic goals of a human being, such as the preservation of one's life.

Extract (15), which was already introduced in Chapter 2, illustrates an example of negation which produces a humorous effect, but where there is no contradiction.

(15) Sharing a tent with a man who was crazy wasn't easy, but Nately didn't care. He was crazy, too, and had gone every free day to work on the officers' club that Yossarian had not helped build. Actually, there were many officers' clubs that Yossarian had not helped build, but he was proudest of the one on Pianosa. It was a sturdy and complex monument to his powers of determination. Yossarian never went there to help until it was finished—then he went there often, so

pleased was he with the large, fine, rambling shingled building. It was truly a splendid structure, and Yossarian throbbed with a mighty sense of accomplishment each time he gazed at it and reflected that none of the work that had gone into it was his. (p. 28)

This extract presents a similar type of conflict to the one discussed with regard to example (14). In both cases, we can talk about conflicting goals related to incompatible societal and personal MOPs. Thus, we can postulate a societal MOP M-HELP BUILD OFFICERS' CLUB and a personal MOP M-REFUSE TO CO-OPERATE, where the former has the goal "participate actively," and the latter "do nothing." Now, the situation is further complicated by the fact that Yossarian is proud of his attitude. This defeats the standard assumption held by a reader regarding the kinds of things one is usually proud of. People are usually proud of things they have done, less typically of things they have not done. Thus, it can be argued that the negative form introduces a defeated expectation, which creates the context where these MOPs gather. The expectation that is not verbalized, that is denied linguistically, would be realized by the affirmative. This conforms to the pattern proposed by Fillmore (1982, 1985) where the opposition between negative and affirmative can be accounted for as a frame contrast. Both the affirmative and the negative can form a meta-MOP PRIDE where the prototype and the expectations usually associated with it are not fulfilled in the reading process. This can be represented as shown in Table 4.3.

Thus, when reading extract (15), a reader's prototypical notion of pride and the expectations associated with it are not fulfilled. This analysis, however, does not explain why a reader perceives this extract as funny and why it is perceived to be informative. Here, as in other examples where negation can be schema refreshing but is not involved in contradiction, we can account for the humorous effect of the extract by establishing an analogy with a similar experience. Schank's (1982) notion of TOP is useful for this purpose.[11] A TOP is defined as a high level structure that establishes connections between apparently unconnected schemata. In this view, extract (14) may remind us of the people who have been proud not to have cooperated with invaders of their countries during the occupation of their country by the enemy. Thus, we can establish a TOP in the terms shown in Table 4.4.

In this table, a connection is established between two war situations, one where an officers' club is built, another where cooperation with the enemy takes place. The goal is an achievement goal, which is to resist pressure to cooperate; the

TABLE 4.3.
meta-MOP: Pride

Prototype:	one is proud of things one does, one's work, one's friends, one's social skills, etc.
Expectation:	with regard to M-Build officers' club: be willing to cooperate.

TABLE 4.4.

Reminding	Goal	Conditions	Features
Building club Cooperating with enemy	Achievement goal	Negative action; uncooperativeness, pride	War situation

conditions are negative action, or that cooperation does not take place, and pride. Again, we need the evaluative component in order to establish a full connection between the schemata. By analogy with the negative character of cooperating with the enemy, cooperating in the building of an officers' club is also perceived as negative. However, the connections establish a striking analogy between the enemy and the higher officers, a parallelism that is recurrent in the novel and is explicitly pointed out by Yossarian and other characters.

(16) The enemy is anybody who is going to get you killed, no matter what side he's on.

The analogy described in these terms can also help us to account for the humorous character of the description. Thus, although cooperation with the enemy is a very serious matter, or, to put it in other words, it is something important, building an officers' club in comparison is a trivial matter. This reveals a process, which is repeated on other occasions in the novel, where trivial situations reveal a more dramatic background.

From the point of view of Cook's (1994) framework, we can interpret the phenomenon as an example where the use of the negative in order to defeat a standardly held assumption based on a prototype of pride, is schema refreshing. The process is understood by the reader by establishing an analogy with a similar situation, which reveals the higher level meaning of the apparent uninformativity of the passage.

It is obvious from this analysis that the schema theoretic model applied must take account of a hierarchy of schemata and a distinction between lower level schemata that provide information about particular scenes and scripts, and higher levels that organize the lower ones.

4.6. CONCLUSION

In this chapter, we discuss the possibilities of applying schema theoretic models to the analysis of negation and humor, in particular, within the context of a literary work. The advantages of some schema theoretic models, such as Fillmore's (1982, 1985), Shanon's (1981), and Pagano's (1994), which provide a point of departure for the analysis of negation as activating frame knowledge are considered. These frameworks are complemented by two other more complex works, Schank's and Abelson's (1977) and Schank's (1982), to be able to account for hierarchical relations of schemata and unexpected variations in schematic packages.

The most obvious advantage of using schema theories to account for negation is the possibility of making reference to different levels of processing, where failure of understanding at the lower levels can be compensated by understanding at a higher level. Lower level failure in the examples discussed in this chapter have to do mainly with the presence of contradictory utterances or with the defeat of expectations to evoke unusual situations. This phenomenon leads to a challenge of commonly held assumptions regarding situations such as war and behavior in war, thus producing the effect that Cook (1994) describes as schema refreshment.

NOTES

1. See also the discussion of negation as subworld in Werth's (1995c) framework in Chapter 3. Here, negation is accounted for as a subworld that changes the parameters established in the text world. Typically, the negative subworld also defeats expectations that have been created by means of the activation of frame knowledge in the discourse situation.

2. Some of the goals (like "sex") may belong to different categories, such as *satisfaction* and *enjoyment*. It can be questioned which of the functions is primary. Furthermore, the classification need not represent universal needs, rather, it seems to reflect the goals of a particular group (white, middle-aged, Western men), as has been pointed out by several authors. The classification should be understood as a guideline rather than a prescriptive classification.

3. For a different approach to schema change and learning, see Rumelhart (1980).

4. Schank (1982) uses the term trip to the dentist, but it seems more adequate to talk about visits to doctors.

5. See Cook (1994) and Semino (1994) for an application of Schank and Abelson's (1977) model and Schank's (1982) framework. Cook uses the frameworks as the point of departure of his theory of schema refreshment and cognitive change. Semino (1994) incorporates Schank's (1982) framework to the interpretation of poetry.

6. See Hall (1996) for a criticism of the notion of schema refreshment. According to Hall, Cook's (1994) framework has little to contribute to the Russian formalists' notion of *defamiliarization*.

7. Schank (1982) also defends a view where schema change operates in the connections between different levels of schemata in a hierarchy.

8. For further well-known discussions of humor theories applied to the interpretation of jokes see Freud (1966), Sacks (1974), Chapman and Foot (1976), and Norrick (1993). Sacks concentrates on the narrative aspects of a joke structure in a conversation. Norrick (1993) also deals extensively with conversational joking, by means of an analysis based on discourse-pragmatic principles.

9. See Nash (1985) for an approach to humor in literature based on stylistic principles. Nash devotes a chapter to the relation between faulty logic, nonsense, and humor, and makes a brief mention of the significance of these aspects in *Catch-22*.

10. See Raskin (1985) for a theory of humor based on the notion of the simultaneous existence of two incompatible scripts.

11. See Semino (1994, 1997) for an application of Schank's (1982) notions of MOP and TOP to the interpretation of schematic relations in poetic worlds.

5

A Text World Model of Negation In Discourse: An Analysis of Joseph Heller's Novel Catch-22

> Catch-22 did not exist, he was positive of that, but it made no difference.
> What did matter was that everyone thought it existed, and that was much worse,
> for there was no object or text to ridicule or refute, to accuse, criticize, attack,
> amend, hate, revile, spit at, rip to shreds, trample upon or burn up.
>
> (Heller, 1961, p. 516)

5.1. INTRODUCTION

In this chapter, an integrated approach to negation in discourse is proposed that brings together some of the semantic, pragmatic, and cognitive features of negation discussed in previous chapters. The general framework adopted for this purpose is based on Werth's (1995c/1999) text world theory, which is complemented and expanded by integrating notions such as Givón's ontology of negative states and events,[1] the idea of complex schematic networks proposed by Schank (1982),[2] and Ryan's (1991b) notion of conflict in fictional worlds.[3] Ryan's (1991b) notion of conflict in the fictional world, in particular, contributes to the linguistic analysis by establishing a link between the linguistic notion of negation as subworld and its function in discourse, on the one hand, and the literary notion of conflict within a fictional world, on the other. The integration of the two approaches can provide insights regarding the ultimate function of negation from the perspective of the processing and understanding of a literary work.

The model proposed is applied to the discussion of relevant extracts from the novel *Catch-22*. The aim of the discussion of this chapter is twofold: (1) to put forward a discourse model of negation; and, (2) to illustrate its application to the analysis of negation within the specific context of the novel *Catch-22*. The latter point involves the development of an interpretation of the recursion of negation within the context of the novel, in particular, by arguing that negation plays a crucial role in the development of a specific world view. The idea is to explore both what the theory has to contribute to the understanding of the novel as a whole, and what the data can contribute to an evaluation of the theoretical framework as a model of negation in discourse. More specifically, the discussion has the aim of exploring the following issues.

1. To propose modifications to the frameworks applied in order to account for idiosyncratic aspects of the data.
2. To identify recurring patterns of negation throughout the novel that can be recognized as specific discourse functions of negation within a text world theoretical model.
3. To point out those cases where specific linguistic functions of negation can be said to be marked and, thus, to contribute to the creation of a defamiliarizing effect and the triggering of schema refreshment.
4. To identify the relation between the function of negation as subworld and the development of patterns of conflict within the fictional world of *Catch-22*.
5. To describe the function of lexical negation and its contribution to the creation of patterns of contradiction.

In this analysis, we focus on the function of clauses in which syntactic negation has an assertive communicative function[4] and on contradiction created both by means of syntactic and lexical negation. With regard to negative assertion or negative statement, I argue that this function is part of the more general function of negation as a term in the polarity system with properties that differentiate it from the affirmative. In Halliday's (1994) terms, we may say that this function is primarily ideational. The communicative function of negative assertion, however, has to be distinguished from other speech acts or communicative functions performed by negative clauses, where the interpersonal component, rather than the ideational one, is primary. This is the case of commissives and directives. For reasons of space, only passing remarks are made on this second type of communicative function of negative clauses. We concentrate on the cognitive, ontological, and textual properties of negative assertion.[5]

5.2. A DEFINITION OF NEGATION

As discussed in Chapter 3, a text world approach to negation takes as a point of departure a series of theoretical assumptions that can be summarized as follows. From a discourse-pragmatic perspective, negation is a marked option that operates in discourse on the assumption that the affirmative term is expected or familiar to speaker and hearer. This is what Givón (1978, 1984, 1993) calls the *presuppositional nature* of negation. Nonevents and nonstates, expressed linguistically by means of negatives, are cognitively less salient than events or states, expressed linguistically by means of the affirmative. This makes negation typically less informative than affirmation. As a consequence, negation is a natural foregrounding device used when, more rarely in discourse, nonevents or nonstates are considered to be more informative than events or states.[6] This argument justifies the view of negation as typically expressed by an illocutionary act of denial, where the denied proposition corresponds to an assumption or expectation implicitly or explicitly present in the common ground of the discourse.

In text world theoretical terms, negation in discourse can be specified further as a subworld; as such, it is defined as follows.

1. As a cognitive or conceptual space, a negative subworld is a nonfactual domain triggered by a negative word, and which describes a nonevent or a nonstate; or, following Langacker, negation can also be said to introduce a mental space where a given event or state does not take place or where a given entity fails to occur (Langacker, 1991, pp. 133–134). The notion of negation as a subworld involves a cognitive reinterpretation of its semantic features; this allows us to account for the fact that negation not only establishes semantic scope over a stretch of discourse, but it also introduces a state of affairs that contrasts with a previously assumed domain, typically, the status quo of the text actual world. In this definition of subworld, negation takes scope over a stretch of discourse within a clause as is traditionally understood.

2. The notion of subworld proposed by Werth (1995c/1999) can be taken further to stand for a discourse unit that is under the influence of one or more negative clauses. In this meaning, a negative subworld has the characteristics of what Martin (1992, p. 553) calls *prosody*, which is described as a phenomenon where "non-discrete realisations, . . . smear across rather than mapping onto elements of schematic structure." Louw (1993, p. 157) applies the notion of *semantic prosody*, which is defined as "A consistent aura of meaning with which a form is imbued by its collocates" to the description of irony in text. In the discussion that follows, I argue that certain episodes illustrate negative subworlds that are semantic prosodies and, consequently, can be defined as discourse units, instead of being defined only as clausal units.

With regard to the functions performed by negative subworlds in discourse, following Werth (1995c, p. 388), it can be said that negation contributes to the general discourse function of up-dating the information of the text world. As such, in Werth's work (pp. 384–385) negative subworlds perform either of two functions.

1. A function of up-dating information in the text world by means of modifying information in the text world, either world-building elements (deictic or subworld-building parameters) or function-advancing propositions (narrative or descriptive in fictional writing).
2. A function whereby an item is both presented and denied at the same time, namely, accommodation.

It can be argued that these functions have to do with the rechanneling or reorientation of ongoing discourse; however, I argue that, in the data under consideration, further categories can be established that are based on different properties of the subworld. Certain types of contradictions do not fit the functional description proposed in these terms, and, for this reason, I propose an account of this type of contradiction as a discourse function of negation that blocks the flow of information. The result is an apparent communicative short circuit, which, however, is resolved by means of inferencing procedures by the reader at a higher processing level.[7] This higher-level resolution of conflict at the textual level also takes place with other instances of marked uses of negation.[8]

To sum up, the text world approach to negation presented here encompasses pragmatic and semantic principles with a strong cognitive basis. This combination enables us to account for the contextual reasons that determine the choice of negation in discourse and to describe the specific functions carried out by negative clauses. The types of functions of negative statements are summarized in Figure 5.1.

In Figure 5.1., the various types of functions of negative statements are specified for those negatives that rechannel the flow of information; however, it is understood that the same categories can be applied to the analysis of pure contradictions because the negative clause will contradict a previous proposition.[9] The term *pure contradiction* is used to establish a difference between those contradictions that can be solved by, for example, ultimately favoring one of the contradictory terms, as is proposed by Sperber and Wilson (Sperber & Wilson, 1986, p. 115), and contradictions that require the simultaneous validity of the contradictory terms. The latter are defined as *pure contradictions*. We return to this point in the discussion of contradictions in Section 5.4.2.

It can be observed that lexical negation is not contemplated as one of the functions of negation in discourse as outlined in Figure 5.1. This is because lexical negation, as indicated in Chapter 2, must be distinguished from what is standardly known as syntactic negation. We can only say a clause has negative polarity when it displays syntactic negation but not lexical negation. In Chapter 2, lexical negation is defined as the phenomenon by which certain words are classified as having a negative value, which is established by opposition to another term typically classified as positive

---

The Discourse Functions of Negative Statements:
A Text World Model for Narrative Discourse

1. General function of rechanneling information in discourse
   1.1. Modification of function-advancing propositions
        Modification of narrative propositions
        Modification of descriptive propositions (negative individuation)
        Modification of information evoked or inferred from function-advancing propositions
   1.2. Modification of world-building parameters
        Modification of deictic information
        Modification of subworlds
        Negative accommodation (negative identification)
2. General function of blocking information in discourse
   Pure contradictions

FIGURE 5.1.

(for example, crazy–sane).[10] This type of opposition is also relevant for the analysis of negation in discourse, and, as such, of *Catch-22*, as is shown in the discussion in Section 5.5.

Before starting with the discussion of the different functions of negative clauses and their significance in the context of the novel *Catch-22*, I illustrate the points made with regard to the features of negative subworlds and their function in discourse by discussing an extract from the novel.

Example (1), which was already discussed in more general terms in Chapter 2, illustrates the denial of information that is inferred from the previous function-advancing discourse; the denials defeat expectations that have been created regarding how a particular mission will take place.

(1) He gasped in utter amazement at the fantastic sight of the twelve flights of planes organized calmly into exact formation. The scene was too unexpected to be true. There were no planes spurting ahead with wounded, none lagging behind with damage. No distress flares smoked in the sky. No ship was missing but his own. For an instant he was paralyzed with a sensation of madness. Then he understood and almost wept at the irony. The explanation was simple: clouds had covered the target before the planes could bomb it, and the mission to Bologna was still to be flown. He was wrong. There had been no clouds. Bologna had been bombed. Bologna was a milk run. There had been no flak there at all. (p. 186)

The unexpectedness of the situation is expressed explicitly: *The scene was too unexpected to be true*. The negative clauses defeat expectations that are associated with the knowledge frame AIR MISSION, such as the fact that there will be damaged planes and wounded soldiers (*There were no planes spurting ahead with wounded, none lagging behind with damage*) and signals of distress (*No distress flares smoked in the sky*). An explanation for the situation is then provided (*clouds had covered the target before the planes could bomb it, and the mission to Bologna was still to be flown*), which is proved to be wrong. Again, the defeated assumptions are coded in the form of negative clauses: *There had been no clouds* and *There had been no flak there at all*. The negative clauses support the main item of information, coded in the form of two affirmative clauses, *Bologna had been bombed. Bologna was a milk run.* In text world theoretical terms, the first negative subworlds (*There were no planes spurting ahead with wounded, none lagging behind with damage, No distress flares smoked in the sky*) deny information that is inferred from function-advancing material and the associated knowledge frames. The proposition *There had been no clouds* modifies descriptive material introduced by the proposition *clouds had covered the target before the planes could bomb it*. Finally, the proposition *There had been no flak there at all* is similar to the negative clauses in the first part of the extract, in that it denies an expectation that arises from previous discourse and the evocation of the relevant knowledge frames.

In general terms, this extract illustrates the way in which negation can modify ongoing discourse and redirect it. It also illustrates the fact that negative clauses typically work together with affirmative clauses and not in isolation.[11]

5.3. NEGATION AS SUBWORLD THAT RECHANNELS INFORMATION

The function of up-dating or rechanneling information is carried out by modifying information that is present in the common ground in the form of an explicit or an implicit proposition (expectation, assumption, or world knowledge).[12] A first distinction is established between negation that modifies function-advancing propositions and negation that modifies world-building information. Although the line that divides function-advancing propositions from world-building material is not clear cut, as argued in Chapter 3, it is used as a basic guideline in the discussion.

5.3.1. Modification of Function-Advancing Propositions

Negative subworlds can modify either narrative propositions or descriptive propositions, or information that is evoked or inferred by either. Each is considered in turn. The distinction between narrative and descriptive propositions is based primarily on the type of predicate involved; thus, narrative propositions are characterized by dynamic predicates and descriptive propositions are characterized by stative predicates (for a detailed discussion, see Chapter 3).

Extracts (2), (3), and (5) are typical examples of the modification of information that arises from the narrative discourse.

(2)   After he had made up his mind to spend the rest of the war in the hospital, Yossarian wrote letters to everyone he knew saying he was in the hospital but never mentioning why. One day he had a better idea. To everyone he knew he wrote he was going on a very dangerous mission. "They asked for volunteers, but someone has to do it. I'll write you the instant I get back." And he had not written anyone since. (p. 14)

In extract (2), we find the two following negative clauses: (1) *but never mentioning why*, and (2) *And he had not written anyone since*. Both clauses can be said to be prototypical examples of the function of negation as up-dating and rechanneling information, as the reader may be holding a wrong assumption or expectation about how the discourse will develop. More particularly, the negative clauses modify expectations that have arisen from narrative function-advancing propositions in the previous discourse. Thus, if we are told that Yossarian is writing letters from the hospital, an expectation is created that he should explain why he is in the hospital. Similarly, if we are told that Yossarian writes that he will write back soon, we assume this to be true. The negative clauses deny the applicability of the assumptions and expectations, so that what would generally be the case, that one usually explains why he or she is in the hospital, that one usually keeps one's word, is denied. This has the effect of foregrounding the negative term, the nonexplanation, the nonreply. The negatives, which manifest a deliberate refusal to provide an explanation of why he is in hospital and a deliberate failure to keep his word, can be understood as provocative moves directed at producing anxiety and worry in the friends who receive the letters and who do not know if Yossarian is ill or not, or even whether he is alive or dead.

In extract (3), negation defeats an expectation regarding what should be predicted to happen according to general assumptions about cause–effect. The extract refers to one of the soldiers who shares the hospital ward where Yossarian is during one of his stays at the hospital.

(3)   With them this time was the twenty-four-year-old fighter-pilot captain with the sparse golden moustache who had been shot into the Adriatic Sea in midwinter and not even caught cold. Now the summer was upon them, the captain had not been shot down, and he said he had the grippe. (p. 15)

In a typical process that is recurrent in the novel, cause–effect relations as we normally understand them are revealed to be inapplicable; negation in extract (3) denies the expectation that a certain effect should follow a cause: if one falls into the sea in midwinter, it is highly probable that one will catch a cold. This expectation is defeated by means of the negative clause *and not even caught cold*. The effect is reinforced by the parallel structure that follows, where the negative event, expressed in the clause *the captain had not been shot down*, describes the absence of the cause that produces the (now) unexpected effect (*he said he had the grippe*). It can be argued that episodes of this kind trigger a specific type of frame knowledge that has to do with logical relations. The negative terms contribute to the undermining of the logical cause–effect structure.

(4)   a.   Cause *x*—Effect *y*
      b.   Cause *x*—Not-Effect *y*
      c.   Not-cause *x*—Effect *y*

As illustrated in (4), although (4)a shows the expected pattern, (4)b and (4)c are variations where the presence of a negative term modifies the pattern, thus defeating an underlying expectation.

To end the discussion of negative subworlds that modify dynamic predicates, we discuss an extract from a dialogue that is part of a trial against a soldier called Clevinger. This is one of the few examples discussed that is not a negative statement but a negative question.

The higher officers proceed by questioning Clevinger on things he has not said or done.

(5)   "When didn't you say we couldn't punish you? Don't you understand my question?"
      "No, sir. I don't understand."
      "You've just told us that. Now suppose you answer my question."
      "But how can I answer it?"
      "That's another question you're asking me."
      "I'm sorry, sir. But I don't know how to answer it. I never said you couldn't punish me."
      "Now you're telling us when you did say it. I'm asking you to tell us when you didn't say it."
      Clevinger took a deep breath. "I always didn't say you couldn't punish me, sir."
      (p. 101)

The question *When didn't you say we couldn't punish you?* illustrates an example of a negative clause, *we couldn't punish you*, embedded within another negative clause, *When didn't you say...* The former is not remarkable in any sense and it illustrates the modification of a modal predicate (*could*) by means of the negative subworld. The latter is a marked case of negation where the negative particle *not* takes scope over the predicate where the main verb is *say*. In communicative terms, the question is difficult to process and to understand. The reasons for the difficulty in understanding the question are several. A *when-* question is an open-ended question that requires specification regarding the particular moment or period in time in which some activity took place. Negation, however, lacks the necessary specificity to describe a particular moment or period in time, unless the range upon which the negative operates is provided by other means. In other words, a negative clause is typically uninformative, as it describes the general background norm upon which events stand out as salient. The number of times when Clevinger did not say they could not punish him is infinite.[13] Consequently, a question that requests information about the background norm is uninformative and apparently pointless.

Furthermore, it can be observed that the *wh-* element falls within the scope of negation, following the tendency, mentioned by several authors,[14] for adjuncts to attract the scope of negation. In this respect, Kuno (1993b, p. 4) argues that a question such as *When didn't you say we couldn't punish you?* violates the *ban on extraction of the target of negation* and the *ban on questions that solicit uninformative answers*. The former establishes that:

> An element that is the Target of Negation cannot be extracted out of the scope (i.e., the C-command domain) of the negative element. . . . When negated sentences involve adjuncts that may or may not be the targets of negation, the semantic contrast is much clearer. . . . The ban applies only to adjuncts in negative clauses which are the targets of negation. (1992b, pp. 4–7)

In such cases, it is argued that negation creates a *negative island* from which an adjunct cannot be extracted (Kuno, 1993a, p. 1). The reasons are both structural and pragmatic, as the question may be paraphrased by a question, *What is the moment in time when you did not say that?* A question of this type would be acceptable only when it is presupposed that there is one single moment when the action was not done. In this case, the negative becomes more informative than the affirmative.

The *ban on questions that solicit uninformative answers* requires not to ask questions that "have excessively many correct answers" (Kuno, 1992a, p. 14). From a communicative point of view, it would make more sense to ask about things Clevinger has said or done. The situation is taken to an extreme when the officer obliges Clevinger to answer in the same terms posed by the question, that is, by expressing explicitly when something *was not said*. The expression *I never said*, is taken to be affirmative, as the verb does not carry the negative marker.

The use of the negative in this extract is bewildering, and it has the function of foregrounding the negative event in a critical situation, that of a trial. This may challenge the reader's assumptions about the fairness of legal procedures in trials as a reflection of the situation described in this scene. The officers are not interested in the truth, in whether Clevinger is guilty or not, they are only interested in proving he is guilty, at whatever cost. This is stated explicitly at the beginning and the end of the trial.

Since most episodes show a mixture of denials of narrative and descriptive material, we turn now to the discussion of extracts that are basically descriptive but may also have narrative function-advancing propositions.

Negation that modifies descriptive function-advancing propositions can be distinguished into *identification* and *individuation* (Werth, 1995c, p. 305). As explained in Chapter 3, *identification* is the world-building process by means of which an entity is introduced in the discourse. This means that negative identification is what Werth defines as *negative accommodation*. *Individuation* is a descriptive function that consists of providing further information about entities that are already present in the discourse. Finally, it should be remembered that there is a further function of *framing*, which has to do with the further addition of information that is evoked from memory. In fact, the framing process takes place constantly throughout the reading process, as has already been shown in the discussion of extract (1).

In the discussion of the modification of descriptive material in an individuating function, we focus on negative clauses that by means of foregrounding negative events and states, contribute to the creation of a conflict between characters' domains and the status quo of the text actual world. This type of conflict is based on games of deceit and pretense, two important phenomena in understanding *Catch-22*. From this perspective, we can classify negative clauses into two main groups: (1) one into which negation denies descriptive information regarding certain states of affairs that have been presented as facts in the text world, and that are now revealed as false; this is a case where something that was described as being true is now revealed as being false; and (2) another where what is denied is descriptive information regarding states of affairs that are presented as true but are taken to be false, even if there might be evidence proving they are true; this has the striking effect of denying the truth of what is real or actual. Both types are related to general strategies of deceit and pretense on the part of some characters, or, in more general terms, they are related to a conflict between characters' domains (knowledge about the status quo in the text actual world, wishes, beliefs, etc.) and the status quo of the text actual world. We review with each type in turn.

Extract (6) is an example of a situation where something that is described as a fact is then presented as false. The negative clauses illustrate the function of a negative subworld as a modifier of descriptive information that has been introduced in the discourse. Extract (6) is part of a longer description of Colonel Cathcart's stays at his farmhouse.

(6)   Officers' clubs everywhere pulsated with blurred but knowing accounts of lavish, hushed-up drinking and sex orgies there and of secret, intimate nights of ecstasy with the most beautiful, the most tantalizing, the most readily aroused and most easily satisfied Italian courtesans, film actresses, models and countesses. No such private nights of ecstasy or hushed-up drinking and sex orgies ever occurred. They might have occurred if either General Dreedle or General Peckem had once evinced an interest in taking part in orgies with him, but neither ever did, and the colonel was certainly not going to waste his time and energy making love to beautiful women unless there was something in it for him. The colonel dreaded his dank lonely nights at his farmhouse and the dull, uneventful days. He had much more fun back at Group, browbeating everyone he wasn't afraid of. However, as Colonel Korn kept reminding him, there was not much glamour in having a farmhouse in the hills if he never used it. (pp. 268–269)

The negative clause *No such private nights of ecstasy or hushed-up drinking and sex orgies ever occurred*, denies the assumption that the propositions in the preceding descriptive scene are true. The negative proposition triggers a nonfactual domain or subworld, which in conceptual terms can be identified with the appearance or illusion that is created by the Colonel. The negative statement is reinforced by other negative clauses that stress the split between the facts and the illusion. Thus, we are told that there could have been orgies but *neither of the two generals* in the squadron is interested in them, nor is Colonel Cathcart. This Colonel, however, is lured into going to the farmhouse periodically so as to keep up the illusion of the orgiastic weekends, as *there is no glamour if one does not make use of the house*. This is one of the typical illustrations where an appearance is set up in order to project an illusory domain that renders a previous assumption or assertion false.

The particular function of this illusion can be adequately explained in terms of Ryan's (1991b) model applied to conflict in fiction. Thus, we can understand this episode as a conflict between opposing wishes within Colonel Cathcart's personal domain. On the one hand, there is a wish to be admired and accepted, which leads him to play the game of pretending to take part in the orgies. This conflicts with his distaste for the activity in itself, so that he has to pretend that those things take place. The consequence is a deceitful state of affairs projected against the facts of the text actual world, which only the reader knows are different. The extract as a whole can be said to induce schema refreshment because it directly reveals deceitful behavior in the military officers, thus making the reader question the reliability of people belonging to this institution.

The questioning of military operations is a recurrent theme and is particularly disturbing when the reader gets to know that the military apparatus continues even if the enemy is no longer there. This can be seen, for example, in extract (7), where Yossarian wishes he could make use of a machine gun instead of being inside a fighter-bomber plane.

(7) Actually, there was not much he could do with that powerful machine gun except load it and test-fire a few rounds. It was no more use to him than the bombsight. He could really cut loose with it against attacking German fighters, but there were no German fighters any more, and he could not even swing it all the way around into the helpless faces of pilots like Huple and Dobbs and order them back down carefully to the ground, as he had once ordered Kid Sampson back down... (p. 419)

The most striking point in extract (7) is that a negative clause (*but there were no more German fighters any more*) corrects a basic assumption up to this point held by the reader, namely, that there was an enemy. The denial of the assumption that there is an enemy leads to a questioning of what the Americans are doing there if there are no enemies left. The rest of the negative clauses (*there was not much he could do with that powerful machine gun; It was no more use to him than the bombsight; he could not even swing it all the way around*) modify assumptions that arise from the description of the scene (there was not much he could do, it was no more use to him) and the assumption that he might actually *do* something with the machine-gun (he could not even swing it round).

The manipulation of facts to create an illusory and deceitful appearance becomes more threatening towards the end of the novel, where the arbitrariness of power in the hands of the higher officers becomes more obvious and more dangerous. The following extract describes how Yossarian finds out that the famous catch-22 does not exist, and that the higher officers use it as an excuse to carry out all sorts of unjust actions.

(8)   Yossarian left money in the old woman's lap—it was odd how many wrongs leaving money seemed to right—and strode out of the apartment, cursing Catch-22 vehemently as he descended the stairs, even though he knew there was no such thing. Catch-22 did not exist, he was positive of that, but it made no difference. What did matter was that everyone thought it existed, and that was much worse, for there was no object or text to ridicule or refute, to accuse, criticize, attack, amend, hate, revile, spit at, rip to shreds, trample upon or burn up. (p. 516)

Negation in this extract carries out an individuating function, in the sense that it denies information regarding a basic property of an entity previously introduced in the text world, the catch-22. Quite surprisingly, the information that is denied is the existence of the catch-22 itself as an accepted entity in the fictional world. In this extract, something that was introduced as real, the catch-22, is revealed to be an illusion, so the truth regarding its existence is denied. This shows a deeper conflict that pervades the whole novel, summarized in two crucial sentences: *Catch-22 did not exist, he was positive of that, but it made no difference. What did matter was that everyone thought it existed, and that was much worse, for there was no object or text to ridicule or refute.* The negative clause *Catch-22 did not exist* denies a firmly held assumption. This denial creates a new expectation that things should change now that a given state of affairs has been revealed to be false, however, the next negative clause, *but it made no difference*, again denies this expectation. This is further explained in the affirmative sentence that follows, *What did matter was that everyone thought it did.* The process of foregrounding the nonobject acquires dramatic proportions, as we are told that there is *no object or text to ridicule or refute.* The nonobject is not only taken for an existing object, but, being nonexistent, it lacks a fixed definition, it does not refer to a fixed entity with specific properties in the fictional world. This makes it an easy instrument of arbitrary abuse of power on the part of evil people.

The conflict is similar to those discussed previously. Something that was accepted as real is revealed as illusory to the reader, while the fact that it is still accepted as real by most of the characters in the fictional world is stressed (this is expressed in the second part of the extract). In terms of Ryan's (1991b) model applied to conflict in fiction, we can interpret it as a conflict based on a deceitful state of affairs that leads to an erroneous or false K-world on the part of the characters with no power (the population in general and the soldiers). This is contrasted with the fact that the persons with power (the higher ranking officers) know the truth, thus their K-world corresponds with the state of affairs in the fictional world.

The fact that the persons with power deliberately manipulate facts to their advantage is manifested explicitly toward the last pages of the novel, where the chaplain tells Yossarian he is a hero for defending the colonels' lives from a Nazi assassin. Yossarian tells him that the story is false.

(9) "You know, Yossarian, we're all very proud of you."
"Proud?"
"Yes, of course. For risking your life to stop that Nazi assassin. It was a very noble thing to do."
"What Nazi assassin?"
"The one that came here to murder Colonel Cathcart and Colonel Korn. And you saved them. He might have stabbed you to death as you grappled with him on the balcony. It's a lucky thing you're alive!" Yossarian snickered sardonically when he understood. "That was no Nazi assassin."
"Certainly it was. Colonel Korn said it was."
"That was Nately's girl friend. And she was after me not Colonel Cathcart and Colonel Korn. She's been trying to kill me ever since I broke the news to her that Nately was dead."
"But how could that be?" the chaplain protested in livid and resentful confusion.
"Colonel Cathcart and Colonel Korn both saw him as he ran away. The official report says you stopped a Nazi assassin from killing them."
"Don't believe the official report," Yossarian advised dryly. "It's part of the deal."
(pp. 546–547)

By means of negation in its individuating function, which is basically corrective in nature, the reader comes to know that the officers have invented a deceitful version of an episode where Nately's girlfriend has tried to kill Yossarian. The officers have spread the version that it was a Nazi assassin trying to kill the colonels. Yossarian denies the truth of this statement (*That was no Nazi assassin*) and provides the true version. As in the prior illustrations, the manipulation on the part of the officers induces a deceitful state of affairs that creates false K-worlds in other characters except Yossarian, who knows the truth. Again, this is a strategy of arbitrary display of power with regard to Yossarian, who has no way of proving that their version is not true. As in previous examples, the reader is explicitly told that the higher officers are doing something wrong by falsifying facts, which leads to a critical attitude on the part of the reader towards them.

The exploitation of illusory worlds is also a speciality of Milo, the mess officer. Below is an extract in which Yossarian refuses to continue helping Milo on one of his commercial "missions," even if Milo has offered him and Orr the company of two virgins.

> (10) "To hell with my mission," Yossarian responded indifferently. "And to hell with the syndicate too, even though I do have a share. I don't want any eight-year-old virgins, even if they are half Spanish."
> "I don't blame you. But these eight-year-old virgins are really only thirty-two. And they're not really half Spanish but only one-third Estonian."
> "I don't care for any virgins."
> "And they're not even virgins," Milo continued persuasively. "The one I picked out for you was married for a short time to an elderly schoolteacher who slept with her only on Sundays, so she's really almost as good as new." (p. 297)

In extract (10), as in (9), the illusory nature or falsity of previous discourse is revealed. In this extract, Milo confesses that the girls he has found for Yossarian and Orr are not really virgins, as he had assured them. Little by little he reveals the series of lies he has told them: *And they're not really half Spanish but only one-third Estonian, And they're not even virgins."* The negative clauses modify assumptions that arise from descriptive information about the entities *girls*. As in (9), the correction of previous propositions reveals Milo's deceitful nature. This revelation can lead to a questioning of commercial procedures and of the honesty of professionals in this field. Such a questioning takes place by means of a process of generalization, where the properties said of Milo are applied to commerce in general terms.

The setting up of pretense pervades other professional areas, such as medicine. Below is an example where Yossarian, during one of his stays in hospital, is obliged to pretend he is the dying son of a couple who have come to visit their son, who is already dead.

> (11) "There are some relatives here to see you. Oh, don't worry," he added with a laugh. "Not your relatives. It's the mother, father and brother of that chap who died. They've traveled all the way from New York to see a dying soldier, and you're the handiest one we've got."
> "What are you talking about?" Yossarian asked suspiciously. "I'm not dying."
> "Of course you're dying. We're all dying. Where the devil else do you think you're heading?"(p. 234)

Whereas the extracts commented on previously are an invitation to criticize the behavior of military officers, this extract questions the humanity of doctors in a war situation. The negative clause *Oh, don't worry . . . Not your relatives*, denies the commonsense assumption Yossarian has made that his relatives are there to visit him. Yossarian is later blackmailed into pretending he is the dying soldier, thus putting up a scene where he pretends to be somebody else. The absurdity of the situation is reinforced by the fact that the relatives do not seem to notice he is not their son, or, if they do, they do not mind. As in previous examples, it is the appearance that matters, that the relatives are able to see *some* dying soldier, and not the reality, that it is not their son.

The deceitful nature of the higher officers shows, as already pointed out, an exaggerated desire to be admired and to stand out above the rest at whatever cost. This leads each of the colonels and generals into petty power struggles among themselves. One of the experts in finding ways of standing out and being admired is General Peckem. The following extract refers to the term *bomb pattern*, which he has invented.

> (12) Colonel Scheisskopf was all ears, "What are bomb patterns?" "Bomb patterns?" General Peckem repeated, twinkling with self-satisfied good humor. "A bomb pattern is a term I dreamed up just several weeks ago. It means nothing, but you'd be surprised at how rapidly it's caught on. Why, I've got all sorts of people convinced I think it's important for the bombs to explode close together and make a neat aerial photograph. There's one colonel in Pianosa who's hardly concerned any more with whether he hits the target or not. (p. 411)

This extract already shows the dangers of the higher officers' obsessions with the appearance of things, rather than with the things themselves. This is particularly threatening when the things they are dealing with have to do with war. This problem can be said to be expressed by a description of the properties of an entity in negative terms: *It means nothing, but you'd be surprised at how rapidly it's caught on*. The negative statement corrects the assumption that an entity should have a basic property, to be meaningful. We can say that this is a prototypical example of a marked foregrounding of a negative state: the general defines a term precisely by its lack of informativity. The second part of the sentence reveals why the term is significant, even if it is not informative: it has caught on quickly. Again, it is the predominance of the appearance over the fact that is the focus of attention. This, however, can have disastrous consequences, as another colonel takes it so seriously that he is convinced that the appearance (the bomb pattern on the aerial photograph) is primary with respect to the reality (where the bombs actually fall). The effect of this extract is reinforced by what is said about the village the bombardiers have to bomb on a mission, described in extract (13).

> (13) "They'll be bombing a tiny undefended village, reducing the whole community to rubble. I have it from Wintergreen. Wintergreen's an ex-sergeant now, by the way—that the mission is entirely unnecessary. Its only purpose is to delay German reinforcements at a time when we aren't even planning an offensive. But that's the way things go when you elevate mediocre people to positions of authority." He gestured languidly toward his gigantic map of Italy. "Why, this tiny mountain village is so insignificant that it isn't even there." (p. 412)

The strategy displayed here is similar to the one described for other extracts discussed, as it also relies on favoring the appearance over the reality. Thus, it is said that *the mission is entirely unnecessary* (one of the examples where morphological negation has a clear foregrounding function), that it is a moment when they *are not even planning an offensive*, and that the village to be bombed *is so insignificant that it isn't even there* (meaning it is not on the map). The first negative clause denies the assumption that, in war, action is carried out either because you need to attack or because you need to organize a defense. In this sense, the negative clause modifies an assumption or expectation that arises from the narrative material. The negative clause also denies the subworld-building word *plan*, thus canceling the intentional component. The second negative clause foregrounds a nonstate, by pointing out that the village *is not on the map*; the negative clause thus modifies assumptions or expectations related to the descriptive information in the preceding discourse.

As in the other extracts discussed in this section, the foregrounding of the negative events and states is explained explicitly. The aim of unnecessarily attacking an insignificant village is to delay German reinforcements, which is also a minor objective. The absurdity of the whole enterprise is made evident by the general's words about the plans made by mediocre people. It constitutes a blatant exposure of the inefficiency and ineptitude of the higher officers.

The second type of negative subworld that alters descriptive parameters set up in the text world is used for two purposes. One is that of denying a fact directly observable from the reality of the fictional world. The other is that of denying information that describes states of affairs previously described as true facts but which are taken by some characters to be false or ignored. The following are some representative instances.

As in the examples of the denial of previous discourse, the denial of what is a fact in the text actual world acquires threatening proportions towards the end of the novel. This can be observed in extract (14), where the chaplain is being accused of having forged his own handwriting.

(14) "This isn't your handwriting."
The chaplain blinked rapidly with amazement. "But of course it's my handwriting."
"No it isn't, Chaplain. You're lying again."
"But I just wrote it!" the chaplain cried in exasperation. "You saw me write it."
"That's just it," the major answered bitterly. "I saw you write it. You can't deny that you did write it. A person who'll lie about his own handwriting will lie about anything." (pp. 481–482)

In this extract, we have a denial of a fact that is obviously true for one of the characters (here, the chaplain) and for the reader: *This isn't your handwriting*. The absurdity of the claim that the chaplain's handwriting is not his handwriting is taken to an extreme by the officer who interrogates him, as he argues that the fact that he just wrote the words is evidence in favor of the fact that he is lying. The explanation to this scene is found elsewhere in the novel, where we are told that both Yossarian and Major Major have been forging documents that have been signed with the names of the chaplain and of Washington Irving. What is important here is that the higher officers are more eager to believe that the chaplain is lying about his identity (and his handwriting), than to believe that someone else may have been forging the signatures.

Linguistically, we can explain the absurdity of the utterance *This isn't your handwriting* as being ambiguous, since the reference of the possessive *your* is not clear.[15] It seems that the officers believe that the chaplain is actually Washington Irving, and are expecting him to use the handwriting they have in previous documents signed by the name of Washington Irving. Conversely, they may believe he is the chaplain, but even then they are expecting him to use the handwriting of the other documents where his name has been forged by somebody else.

Indeed, the chaplain himself seems to have great difficulties in interpreting real facts as real, as is shown in his long reflections about déjà vu and other extraordinary perceptual phenomena. He is particularly obsessed with the episode of Snowden's funeral, where he thinks he has a vision of a man sitting on a tree.[16]

(15) There was no mistaking the awesome implications of the chaplain's revelation: it was either an insight of divine origin or a hallucination; he was either blessed or losing his mind. Both prospects filled him with equal fear and depression. It was neither déjà vu, presque vu nor jamais vu. It was possible that there were other vus of which he had never heard and that one of these other vus would explain succinctly the baffling phenomenon of which he had been both a witness and a part; it was even possible that none of what he thought had taken place, really had taken place, that he was dealing with an aberration of memory rather than of perception, that he never really had thought he had seen, that his impression now that he once had thought so was merely the illusion of an illusion, and that he was only now imagining that he had ever once imagined seeing a naked man sitting in a tree at the cemetery. (p. 341)

This extract displays a complex rationalization process where the real is given the proportions of the extraordinary. The chaplain is trying to make sense of the episode where Yossarian is sitting on the branch of a tree during Snowden's funeral. The interpretation of this fact as something extraordinary is indicated from the beginning, where the chaplain tries to identify the "phenomenon" by an elimination process: *it was either an insight of divine origin or a hallucination; he was either blessed or losing his mind.* The system of binary contrasts that mutually exclude each other is a clear example of the tendency to categorize experience in terms of binary oppositions in a relation of contradictoriness, rather than contrariety.[17] This means that even if there are other terms in between the opposite extremes, this is not perceived as such by the chaplain. Quite significantly, one of the options left out, ironically identified as one of *other vus*, is reality itself, the fact that the vision did actually take place. Thus, the chaplain continues his search of extraordinary sensorial perception in his obsession to find an answer to the enigma of the man on the tree: *It was neither déjà vu, presque vu nor jamais vu.*

The sentence introduced by *it was even possible that* introduces a counterfactual domain in which the subsequent that-clauses are embedded. With regard to the embedding of negative clauses within other subworlds, this topic is discussed in detail in Section 5.3.2.1. Here, I point out that the chaplain's perception that the vision may have been real loses ground progressively.

Linguistically, this loss of touch with the perception of facts is reflected in a distancing process by means of an increasingly complex counterfactual domain in which the negative clauses play a crucial role: *it was even possible that none of what he thought had taken place, really had taken place,* and *that he never really had thought he had seen.* The negative clauses carry counterfactuality to an extreme by denying even the possibility of the existence of the fact referred to, which now becomes *the illusion of an illusion.* This self-reflective process is a recurrent image in the novel, similar to the soldier in white's image and the catch-22 itself. All of them constitute closed circular systems that seem to have an independent status from reality.

The fact that the chaplain is unable to identify the episode as real is indicated explicitly.

(16) The possibility that there really had been a naked man in the tree—two men, actually, since the first had been joined shortly by a second man clad in a brown mustache and sinister dark garments from head to toe who bent forward ritualistically along the limb of the tree to offer the first man something to drink from a brown goblet—never crossed the chaplain's mind. (p. 346)

Extract (17) provides a further example of the denial of knowledge evoked or inferred from previous function-advancing propositions in such a way that a fact that has taken place in the text actual world is treated as if it had not taken place.

(17) Everyone in the squadron knew that Kid Sampson's skinny legs had washed up on the wet sand to lie there and rot like a purple twisted wishbone. No one would go to retrieve them, not Gus or Wes or even the men in the mortuary at the hospital; everyone made believe that Kid Sampson's legs were not there, that they had bobbed away south forever on the tide like all of Clevinger and Orr. Now that bad weather had come, almost no one ever sneaked away alone any more to peek through bushes like a pervert at the moldering stumps.

There were no more beautiful days. There were no more easy missions. There was stinging rain and dull, chilling fog, and the men flew at week-long intervals, whenever the weather cleared. (p. 437)

In extract (17), the negative clauses *No one would go to retrieve them, not Gus or Wes or even the men in the mortuary at the hospital* deny the expectation created from the previous discourse (*Everyone in the squadron knew that Kid Sampson's skinny legs had washed up on the wet sand*). Similarly, the clause *almost no one ever sneaked away alone any more to peek through bushes like a pervert at the moldering stumps* also denies an assumption that can be said to be evoked by the previous discourse. The defeat of expectations that constitute what would be considered as "expected behavior" (that is, that a body is recovered from the beach instead of being left to rot) is reinforced by the following clause: *everyone made believe that Kid Sampson's legs were not there.* The embedding of the negative clause within a cognitive world-building predicate such as *believe* has the effect of showing how a state of affairs that is not the case (Kid Sampson's legs are not on the beach) is taken as if it were true. As in previous examples discussed, there is an open conflict between the status quo in the text actual world and how this status quo is interpreted by the characters and how they react to it. Passage (17) is, perhaps, particularly significant as it indicates the end of a part of the novel, that part where there was still some hope left in the characters. The loss of hope is expressed metaphorically by means of negative clauses that describe how the climate has suddenly changed: *There were no more beautiful days. There were no more easy missions.* These negatives seem to apply in a more general sense to previous discourse by denying the whole way in which events have taken place so far. From that moment on, things will no longer be the same.

To end this section, we briefly consider one of the symbolic images of the novel, which appears strategically at different points. It has to do with the figure of the soldier in white, a soldier who is entirely covered with plaster, so that he cannot move or talk. The soldier apparently dies at some point in the story, but he comes back, causing panic in the hospital ward. This leads Dunbar, one of the soldiers, to think that there is actually nobody inside the bandages, that the plaster case is empty.

(18) There's no one inside!" Dunbar yelled out at him unexpectedly. Yossarian felt his heart skip a beat and his legs grow weak. "What are you talking about?" he shouted with dread, stunned by the haggard, sparking anguish in Dunbar's eyes and by his crazed look of wild shock and horror. "Are you nuts or something? What the hell do you mean there's no one inside?"

"They've stolen him away!" Dunbar shouted back. "He's hollow inside, like a chocolate soldier. They took him away and left those bandages there."

"Why should they do that?"

"Why should they do anything?" (p. 462)

The negative utterance *There's no one inside!* denies the assumption based on previously introduced information that beneath the plaster cover there is a wounded soldier, thus denying the existence of an entity who has previously been introduced as having the property of being alive. As in previous examples, the negative

TABLE 5.1.

Reminding	Goal	Conditions	Features
Colonel's farmhouse/ Catch-22/ Nazi assassin/ Virgins/ Dying soldier/ Bomb patterns/ Bombing village/ Plane attack/ soldier in white	Achievement goal: deceit; manipulate power	The apparent is more important than the real	War situation; officers' behavior; doctors' behavior;

subworld is performing the individuating function of modifying assumptions based on world-building material. The apparently absurd observation that there might be no one inside the plaster produces terror in the soldiers in the ward, which seems to indicate that the claim may be true.

At this point, all the descriptive extracts mentioned so far are brought together to point out how they form part of a pattern that can be perceived throughout the novel. This pattern can be summarized as one where *what is not* takes over *what is*. In the extracts discussed here, *what is not* is an illusory appearance that is imposed on reality by means of deceit. The means whereby we perceive the different extracts as connected can be accounted for by using Schank's (1982) notion of TOP. A TOP is a high level schema organizing structure that enables us to find similarities between apparently disparate experiences.[18] We can say that each of the situations described in the previous extracts evoke schemata which, although not directly connected, present similarities that make it possible to collect them under one single TOP. This is illustrated in Table 5.1.

Thus, the category labeled *reminding* collects the episodes that are perceived as being connected, in this case, the extracts discussed in the present section. We can observe that the different images or schemata have an underlying *goal*, which shows evil intent, in the sense that its aim is to deceive. The goal is an achievement goal because it has to do with the display and abuse of power. The *condition* is that the appearance is always perceived to be more important that the reality,[19] and the *features* are related to the general war situation depicted in the novel but concentrate on the behavior of people with power, the higher officers, the doctors, and the businessmen.

5.3.1.1. Modification of information evoked from function-advancing propositions

The following group of extracts illustrate denials of narrative and descriptive material that evoke significant cultural frames; this means that the negative propositions are defeating expectations that are associated with such cultural frames. For this reason, these examples of negation are considered to be marked, in the sense that they deny standardly accepted assumptions about how things are in the world. They are examples of schema refreshment, as the challenging of these taken-for-granted assumptions leads the reader to reflect upon the issue that is challenged.

(19) His specialty was alfalfa, and he made a good thing out of not growing any. The government paid him well for every bushel of alfalfa he did not grow. The more alfalfa he did not grow, the more money the government gave him, and he spent every penny he didn't earn on new land to increase the amount of alfalfa he did not produce. Major Major's father worked without rest at not growing alfalfa. On long winter evenings he remained indoors and did not mend harness, and he sprang out of bed at the crack of noon every day just to make certain that the chores would not be done. He invested in land wisely and soon was not growing more alfalfa than any other man in the county. (p. 110)

In extract (19), the negative clauses foreground the nonevent of not growing alfalfa and the related activities and properties in this nonproductive process. The first group of negative clauses (up to *the amount of alfalfa he did not produce*) take part in the descriptive function; whereas the second group are mostly narrative and focus on nonevents. The first negative clause *out of not growing any* denies the expectation that arises from the first clause *his speciality was alfalfa*. The following negative clauses (*for every bushel of alfalfa he did not grow; The more alfalfa he did not grow; every penny he didn't earn*; and *the amount of alfalfa he did not produce*) carry out a descriptive individuating function, in that they describe (negative) properties coded in the form of reduced relative clauses. As already pointed out, it is not usual to provide details about an entity in negative terms; consequently, this phenomenon tends to make negation marked, as in these passages.

The sequence of negative clauses creates a negative discourse subworld that describes a nonfactual state of affairs. We are provided with details regarding how the usual procedure by means of which someone earns money by doing something is substituted by the negative version of earning one's living by *not doing* something. In general terms, this denies the commonly held assumption, which is culturally based, that people earn their living by doing things, and less frequently, by not doing them. In text world terms, the negative propositions project a negative subworld that creates a nonfactual domain by means of denying the applicability of a standardly accepted frame of world knowledge. In Schank's (1982) model, we can interpret the extract as involving a meta-MOP (memory organization packet) Mm-EARN A LIVING, where the prototype and the expectations are not fulfilled.[20] This is shown in Table 5.2.

This analysis confirms the view of negation as defeating an expectation or assumption that is typically associated with the expression of an affirmative

TABLE 5.2.

Mm-MOPMm-EARN A LIVING
Prototype: one earns a living by doing something, by working.
Expectation: to earn a living with alfalfa, one usually plants alfalfa in order to sell it.

proposition. This leads to the establishing of parallels between the expected utterance and its defeated version in the negative, as illustrated in Table 5.3.

The foregrounding of the negative produces a striking relation between the expected affirmative utterance and the expressed negative. In a way, this process can also be described as one of reminding, in Schank's terms, because the negative terms remind us of the affirmative, which is the more usual, the one we typically have stored in memory.[21]

Taken literally, extract (19), like the others of this type, is not informative, as it indulges in a long description of the things a character does not do. At this point, Schank's (1982) notion of TOPs (thematic organization packets) prove useful to explain why uses of negation of this kind are interpreted as meaningful and informative by the reader.[22] In Chapter 4, TOPs are described, following Schank, as high order schematic structures that organize and establish connections between schemata belonging to different domains or fields. It was pointed out that the notion of TOP relies heavily on the idea of reminding, as a TOP brings together structures that recall each other. I argue that this extract is interpreted as meaningful because a similarity can be found with standard practices in economic policies of the developed countries. This practice has been a long-established agricultural policy, so as to prevent disastrous price drops.[23]

TABLE 5.3.

Expectation (expressed typically in the affirmative)	Defeated Expectation (expressed by a negative clause)
He made a good thing out of growing a lot of it.	He made a good thing out of not growing any.
For every bushel of alfalfa he grew.	For every bushel of alfalfa he did not grow.
The more alfalfa he grew. . .	The more alfalfa he did not grow, the more money the government gave him.
And he spent every penny he earned	And he spent every penny he didn't earn on new land
To increase the amount of alfalfa he produced	To increase the amount of alfalfa he did not produce
Major Major's father worked without rest at growing alfalfa	Major Major's father worked without rest at not growing alfalfa

TABLE 5.4.

Reminding	Goal	Conditions	Features
Not growing alfalfa/ Uprooting vines	Possession goal	Negative action required	Agricultural policies in developed countries

In Schank's (1982) model, the situation described in (19) is accounted for as a TOP, which establishes a connection between Major Major's father's "occupation" of not growing alfalfa and the agricultural policies followed in developed countries. The structure of the TOP represented in Table 5.4.

By means of this process of reminding, which connects the ideas of not growing alfalfa and being paid for, for example, uprooting vines, and by establishing the analogies between the goals (possession goal = earn money), the conditions (the money is given as a reward for negative action), and features (both have to do with agricultural policies), a reader can interpret the passage as meaningful and informative. The interpretation requires that the reader evaluate the reminded term as negative, that is, that being paid to uproot vines is not good, and that being paid not to grow something, by analogy, rather than absurd, is reformulated as being bad too. This is what leads a reader to question previously held assumptions about how economy works at present in developed Western countries. It is in this sense that negation, as a foregrounded linguistic feature, leads to schema refreshment as defined by Cook (1994).

A similar interpretation can be carried out with regard to extract (20).

(20) Nothing we do in this large department of ours is really very important, and there's never any rush. On the other hand, it is important that we let people know we do a great deal of it. Let me know if you find yourself shorthanded. I've already put in a requisition for two majors, four captains and sixteen lieutenants to give you a hand. While none of the work we do is very important, it is important that we do a great deal of it. Don't you agree? (p. 406)

In this example, we find the following negative clauses: *Nothing we do in this large department of ours is really very important, and there's never any rush; While none of the work we do is very important.* The negative clauses deny expectations that arise from descriptive material and trigger frame knowledge on the theme WORK. It can be argued that the negative clauses project a subworld that denies the applicability of the culturally shared assumption that in work, as in many other things, quality (what we do) is more highly valued than quantity (how much we do). As in example (19), we can interpret the effect of this extract by means of establishing an analogy between the attitude of the higher officer to work and that of another person's. This situation may remind us of the prototypical description

TABLE 5.5.

Reminding	Goal	Conditions	Features
Higher military officers/ Bureaucrats	Achievement goal	Produce unimportant work	Work in offices

of a bureaucrat, that is, a person who works in an office writing and signing lots of papers that have no apparent practical utility. The features that connect the two experiences can be collected by means of a TOP, as shown in Table 5.5. In this case, the shared achievement goal involves producing a great amount of work. The condition is that the work should not be important. Features are the situation in an office environment.

As in example (19), the reminded situation is evaluated negatively, that is, it is bad to give priority to quantity rather than to quality. By analogy, we interpret the extract as a description where such a situation type leads to a negative judgment, that is, it leads to questioning and criticizing behavior of that kind.

Similarly, extract (21) foregrounds the apparent lack of interest in pennants won as prizes in parades.

> (21) To Yossarian, the idea of pennants was absurd. No money went with them, no class privileges. Like Olympic medals and tennis trophies, all they signified was that the owner had done something of no benefit to anyone more capably than anyone else. (p. 95)

In this extract, we identify the following negative clauses: *No money went with them, no class privileges*; and *that the owner had done something of no benefit to anyone*. The subworld described by the negative terms refers to the prestige associated with prizes. Negation denies the commonly held assumption that the winning of prizes is rewarding because it gives status, prestige, and glory, even if it is not economically productive. Yossarian rejects this value system, implying that in his set of priorities, money is on a higher level than status or prestige. In terms of Schank's (1982) model, we can explain this extract as a conflict within a meta-MOP Mm-PRIZES where there are two MOPs with conflicting goals: a societal MOP M-ACHIEVE STATUS, and a personal MOP M-EARN MONEY. Clearly, giving priority to one of the MOPs leads to a negative evaluation of the other, as expressed in the extract.

To end the discussion on negatives that deny material that evokes significant cultural frames, I comment on an extract where syntactic negation and lexical negation both contribute to the creation of a common effect, that of defeating expectations about standard procedures in the world of business. This effect is striking and humorous.

(22) Colonel Cargill, General Peckem's troubleshooter, was a forceful, ruddy man. Before the war he had been an alert, hard-hitting, aggressive marketing executive. He was a very bad marketing executive. Colonel Cargill was so awful a marketing executive that his services were much sought after by firms eager to establish losses for tax purposes. Throughout the civilized world, from Battery Park to Fulton Street, he was known as a dependable man for a fast tax write-off. His prices were high, for failure often did not come easily. He had to start at the top and work his way down, and with sympathetic friends in Washington, losing money was no simple matter. It took months of hard work and careful misplanning. A person misplaced, disorganized, miscalculated, overlooked everything and opened every loophole, and just when he thought he had it made, the government gave him a lake or a forest or an oilfield and spoiled everything. Even with such handicaps, Colonel Cargill could be relied on to run the most prosperous enterprise into the ground. He was a self-made man who owed his lack of success to nobody. (p. 40)

We can identify the following negative clauses: *for failure often did not come easily; losing money was no simple matter*; and *He was a self-made man who owed his lack of success to nobody*. Each of the negative clauses defeats an expectation created by the frame knowledge previously activated in the discourse. Thus, we expect failure to be something easy to achieve, while success is more difficult, and, similarly, losing money should be easy. Finally, the last negative clause is a pun on the expression *to owe your success to somebody*, in the same way as *failure did not come easily* is a pun on the expression with *success* as subject. By making all its terms negative (*lack of success, to nobody*), the idiom is inverted and becomes apparently nonsensical. The negative clauses interact with a high number of negative lexical items which reinforce the effect produced, where a familiar situation is described in detail in a sort of negative mirror image (for example, *bad, misplaced, disorganized, miscalculated, overlooked*, and so on) The negative clauses are good examples of what Givón (1979, 1984) calls the *presuppositional nature* of negation, which refers to the property of negation to evoke the corresponding affirmative term. The humor of the present extract lies precisely in the interface between the expected affirmative term and its negation in the discourse. The effect is striking and humorous for two reasons. First, because it defeats what is expected and it foregrounds the unexpected. Second, it reveals a procedure in the business world that is not at all unusual, although it is not usually described in these terms. As in the previous examples, we can establish an analogy with other familiar situations, which leads to a questioning of the validity and sincerity of well-known businesses and businessmen. This is shown in Table 5.6.

Extract 22 presents several similarities to extract (18), as there is a possession goal with evil intent, to earn money with the condition that it is obtained by negative action, and, implicitly, by deceit. The deceit is not in Colonel Cargill himself, however, but in the persons who deliberately hire his services.

TABLE 5.6.

Reminding	Goal	Condition	Features
Col. Cargill/Dishonest businessmen	Possession goal: earn money	Negative action	Business world

To sum up, we can say that denials of propositions that evoke significant cultural frame knowledge are exploited in *Catch-22* to foreground deceitful procedures in the worlds of business, economy, and work. This is done by means of projecting negative subworlds that create discourse domains that invert or defeat expectations about how things usually take place in society.

5.3.2.   Modification of World-Building Information

As explained in Chapter 3, Werth (1995c, p. 290) distinguishes between participant accessible subworlds (time and place) and character accessible subworlds (epistemic, cognitive and hypothetical). The modification of time and place has been discussed as part of the negation of descriptive material, since usually time and place expressions are realized in the form of clause adjuncts that modify predicates or clauses. In these cases, the negative form is either the negative particle *not*, which is attached to the verb, or the words *never* and *nowhere*. In these cases, we do not talk about time and place subworlds, but, rather, of negative subworlds that take scope over time and place adjuncts.

5.3.2.1.   *Modification of character-accessible subworlds*

In this section, we focus on the modification of character accessible subworlds, that is, negation that denies the applicability of information regarding epistemic, cognitive, or hypothetical domains.

Pagano (1994, p. 264) pointed out that negative clauses introduced by modal verbs seem to constitute a special category of negative function. This author provides the following example.

> (24)   *Saabs may not look large.* Yet the Saab 9000 is the only imported car in the USA rated "large" by the Environmental Protection Agency. (*Business Week International*, 12 March 1990, p. 1)

Pagano (1994, p. 264) argues that by means of the denial, the author is making a concession that *the Saabs actually do not look large*, followed by a *but*. Pagano points out that "The writer admits something but then presents an alternative which reduces the effect of the denial." This phenomenon is easily understood if we have in mind the notions of the presuppositional nature of negation and of the world-building character of modal verbs and certain other predicates. In the case of the example provided by Pagano, we can argue that the combination of the modal *may* together with the negation introduces a property (*the Saab looks large*) in order to deny it. In the novel *Catch-22* there are numerous examples of this kind. Their function should fall in general terms within the broad category previously defined, where the negative rechannels information. The negation of world-building predicates such as these, however, presents idiosyncratic characteristics that makes it different from the rest of the subworlds discussed so far and makes it similar to negative accommodation. The following types of denial of world-building parameters are discussed.

1. Denial of epistemic domains, such as *not + possible, not certain*, etc.
2. Negative hypothetical domains, such as *if... not, had...not*, etc.
3. Denial of cognitive domains, such as *not + know, hope, seem, believe, realize.*

Each is considered in turn. The following is an example of a denial of an epistemic domain introduced by the modal *will*, indicating certainty and projection in the future. The extract describes an episode in which an announcement has been put up postponing the celebration of a parade.

> (25) "What's so different about this Sunday. I want to know?" Hungry Joe was demanding vociferously of Chief White Halfoat. "Why won't we have a parade this Sunday when we don't have a parade every Sunday? Huh?"
> Yossarian worked his way through to the front and let out a long, agonized groan when he read the terse announcement there:
> *Due to circumstances beyond my control, there will be no big parade this Sunday afternoon.*
> *Colonel Scheisskopf.* (p. 403)

In extract (25), the negative term *not* is attached to the modal *will* (*Why won't we have a parade this Sunday; there will be no big parade this Sunday afternoon*). The combination of both terms creates a subworld that has the characteristics of a counterfactual projection in the future. It seems that negative *will* or *won't* presupposes that the action described by the main verb following the modal has taken place before, and that in the future, it will no longer be the case. This is perceived by Hungry Joe on reading the notice, as he does not understand the communicative value of a notice where an activity that has not happened before is presented as if it had. The property of *won't* to presuppose the affirmative is exploited consciously by the higher officers, as is reflected in extract (26).

> (26) "What do you know about?" he asked acidly.
> "Parades," answered Colonel Scheisskopf eagerly, "Will I be able to send out memos about parades?"
> "As long as you don't schedule any." General Peckem returned to his chair still

wearing a frown. "And as long as they don't interfere with your main assignment of recommending that the authority of Special Services be expanded to include combat activities."

"Can I schedule parades and then call them off?"

General Peckem brightened instantly. "Why, that's a wonderful idea! But just send out weekly announcements postponing the parades. Don't even bother to schedule them. That would be infinitely more disconcerting." General Peckem was blossoming spryly with cordiality again. "Yes, Scheisskopf," he said, "I think you've really hit on something. After all, what combat commander could possibly quarrel with us for notifying his men that there won't be a parade that coming Sunday? We'd be merely stating a widely known fact. But the implication is beautiful. Yes, positively beautiful. We're implying that we could schedule a parade if we chose to. I'm going to like you, Scheisskopf. Stop in and introduce yourself to Colonel Cargill and tell him what you're up to. I know you two will like each other." (p. 410)

In this extract, we are shown the process by which General Peckem realizes the advantages of allowing Scheisskopf to postpone parades. In this way, Sheisskopf is satisfied, and so is the General, since putting up notices postponing something implies they *could* take place. General Peckem seems to perceive the presuppositional nature of *won't* and uses it to his own advantage, as a means of displaying power in the face of other generals. As in previous examples, the maneuver is just an appearance, an illusion, but it is certainly extremely effective.

With regard to the projection of negative hypothetical domains, conditionals are used to create counterfactual domains that seem to compete with the status quo of reality. An example is extract (27).

(27) "I really can't believe it," Clevinger exclaimed to Yossarian in a voice rising and falling in protest and wonder. "It's a complete reversion to primitive superstition. They're confusing cause and effect. It makes as much sense as knocking on wood or crossing your fingers. They really believe that we wouldn't have to fly that mission tomorrow if someone would only tiptoe up to the map in the middle of the night and move the bomb line over Bologna. Can you imagine? You and I must be the only rational ones left." In the middle of the night, Yossarian knocked on wood, crossed his fingers, and tiptoed out of his tent to move the bomb line up over Bologna. (p. 156)

In a process typical of the world of *Catch-22*, a possible domain is projected (*if someone would only tiptoe up to the map in the middle of the night and move the bomb line over Bologna*) and an unacceptable consequence is derived (*we wouldn't have to fly that mission tomorrow*). As Clevinger points out, the relation between the hypothesis and its consequence actually inverts the ordinary cause–effect relation between events. Instead of assuming that, once the city is conquered, the bomb line will be moved over the city, the soldiers assume that the inverse process may also be true: if one moves the bomb line over the city, it will be conquered, and, consequently, it will not be necessary to fly the mission. The role of negation in this process is precisely that of creating a nonfactual domain that is the consequence of a premise that does not describe the state of affairs of the text actual world, but only the state of affairs in a pretended domain, that of a map. As in previous examples, there is a conflict between a pretended domain, that created on the map, and reality, what is the case in the fictional world. Even more striking is the fact that Yossarian does move the bomb line over the city on the map, and, as a consequence, he causes everybody to believe the city has been won. This process can be described as a kind of negative determinism where events that are initially presented as *not likely/not possible* (it is not possible for the town to be won only by moving the bomb line over the city on the map) is taken as if it were actual (the city has been won). In other words, we can express it as a process where what cannot be becomes what is. This process provides a different view of the relation between the apparent and the real. The phenomenon is certainly not restricted to the presence of negative clauses; rather, it is found in a combination of different linguistic structures that express possibility and consequence. Significantly, there are several episodes in which we can recognize this pattern, as in the deaths of Hungry Joe and Chief White Halfoat. Hungry Joe dreams every night that a cat is suffocating him, and ends up suffocated by a cat in his sleep. Chief White Halfoat is convinced he will die of pneumonia, and he does.[24]

Yossarian seems to believe in the possibility of inverting cause–effect relations, and of changing the course of events, as is seen in extract (27). This can also be observed in extract (28), about the supposed death of the soldier in white.

(28) Now that Yossarian looked back, it seemed that Nurse Cramer rather than the talkative Texan, had murdered the soldier in white; if she had not read the thermometer and reported what she had found, the soldier in white might still be lying there alive exactly as he had been lying there all along, encased from head to toe in plaster and gauze with both strange, rigid legs elevated from the hips and both strange arms strung up perpendicularly, all four bulky limbs in casts, all four strange, useless limbs hoisted up in the air by taut wire cables and fantastically long lead weights suspended darkly above him. Lying there that way might not have been much of a life, but it was all the life he had, and the decision to terminate it, Yossarian felt, should hardly have been Nurse Cramer's. (pp. 214–215)

In this extract, we have the projection of a negative hypothetical domain which describes a counterfactual state of affairs (*if she had not read the thermometer and reported what she had found*). The consequence, however, is unacceptable (*the soldier in white might still be lying there alive exactly as he had been lying there all along*). It is unacceptable because it is based on the assumption that if a person is not known to be dead by other people, that person cannot be dead. There is an unacceptable cause–effect relation established between the act of perception, of realization of a state of affairs in the actual world that someone is dead, and the result of an independent natural process, the death of a soldier for other reasons which the reader does not know.[25]

The fact that Yossarian does intervene actively in the modification of the course of events is mentioned explicitly in extract (29). This passage summarizes the sudden realization that apparently absurd actions carried out by Yossarian, like forging signatures in the censoring of letters (see extract 2) and moving the bomb line over the map have had dramatic consequences.

> (29) In a way it was all Yossarian's fault, for if he had not moved the bomb line during the Big Siege of Bologna, Major de Coverley might still be around to save him, and if he had not stocked the enlisted men's apartment with girls who had no other place to live, Nately might never have fallen in love with his whore as she sat naked from waist down in the room full of grumpy blackjack players who ignored her. (p. 363)

In this extract, we have two hypothetical negative clauses that project counter-factual domains: *if he had not moved the bomb line during the Big Siege of Bologna*; and *if he had not stocked the enlisted men's apartment with girls*. The consequence is *Nately might never have fallen in love with his whore*. Each of these clauses presupposes a proposition with the contrary truth value, that is, *Yossarian did move the bomb line, Major de Coverley did stock the apartment with girls*, and *Nately did fall in love with the whore*. The ultimate consequence of this process is not mentioned here, but the reader finds out later that Nately decides to stay in Italy instead of going back to the United States and is killed on a mission. Nately's death symbolically represents the loss of the little hope that was left, as virtually all of Yossarian's friends have died or disappeared.

To sum up, negative hypothetical subworlds have the function of projecting counterfactual domains that describe impossible or unlikely states of affairs, from which an unacceptable consequence is inferred. The consequence, however, is accepted as actual and valid by the characters in the world of *Catch-22*. This leads to the creation of further pretended worlds that have dramatic consequences on the development of events in the story.

To end this section, some examples of negation of a cognitive domain are mentioned as they are also relevant for the understanding of the function of negation in the novel. There are several episodes that describe the inability of some characters to determine their knowledge of what is happening around them or to themselves, even in cases where they should be expected to know what is going on. An example is Major Major, about whom it is said:

> (30) He had been made squadron commander but had no idea what he was supposed to do as squadron commander unless all he was supposed to do was forge Washington Irving's name to official documents and listen to the isolated clinks

and thumps of Major de Coverley's horseshoes falling to the ground outside the window of his small office in the rear of the orderly-room tent. (p. 119)

Here, we are told that Major Major *had no idea* what it means to be squadron commander, that is, what his duties are exactly. In particular, Major Major is concerned about his relative status with regard to Major de Coverley, a mysterious and charismatic character who everybody respects without knowing exactly what his rank is.

(31) Major Major wondered about his relationship to Major de Coverley and about Major de Coverley's relationship to him. He knew that Major de Coverley was his executive officer, but he did not know what that meant, and he could not decide whether in Major de Coverley he was blessed with a lenient superior or cursed with a delinquent subordinate. He did not want to ask Sergeant Towser, of whom he was secretly afraid, and there was no one else he could ask, least of all Major de Coverley. Few people ever dared approach Major de Coverley about anything and the only officer foolish enough to pitch one of his horseshoes was stricken the very next day with the worst case of Pianosan crud that Gus or Wes or even Doc Daneeka had ever seen or even heard about. Everyone was positive the disease had been inflicted upon the poor officer in retribution by Major de Coverley, although no one was sure how. (p. 120)

Major Major does not know the status of Major de Coverley, and this is also unknown to the rest of the soldiers, who seem to believe Major de Coverley has extraordinary powers, as is expressed in the last sentence of the extract.

The lack of knowledge of characters with regard to their work is also a characteristic of the doctors of *Catch-22*, as is observed in the following two extracts.

(32) "It's not my business to save lives," Doc Daneeka retorted sullenly.
"What is your business?"
"I don't know what my business is." (p. 224)

In this extract, Doc Daneeka, quite surprisingly, states he does not know what his business is. This goes against general world knowledge regarding the professional and moral commitment of doctors to their work.

Finally, it is the chaplain who has the most difficulties in knowing what is going on, as he seems to be completely unable to differentiate between reality and illusion. This doubt is expressed several times and has already been mentioned in the comments on extract (16). Here are two further examples.

(33) Perhaps he really was Washington Irving, and perhaps he really had been signing Washington Irving's name to those letters he knew nothing about. Such lapses of memory were not uncommon in medical annals, he knew. There was no way of really knowing anything. (p. 339)

The chaplain's doubt about what is real and what is not real is encapsulated in the negative sentence *There was no way of really knowing anything.* This uncertainty partly reflects the general uncertainty about the outcome of events in the world of *Catch-22*, since, as we have seen in previous episodes, it is extremely difficult to make predictions about how things will turn out, as the world of *Catch-22* seems to develop rules of its own. It is also obvious, however, that the chaplain has succumbed to this pattern of uncertainty and is not able to state his own identity. In a typically schizophrenic attitude, he claims the possibility that he might be someone else (Washington Irving). The pressure of the external rules has become too strong for the chaplain, who seems to be on the verge of believing the wrong assumptions others have of him. The uncertainty affects not only his recognition of his own identity and the distinction between real and illusory states of affairs, but also the difference between moral values, what is good and what is bad, what is monstrous and what is not monstrous.

> (32) So many monstrous events were occurring that he was no longer positive which events were monstrous and which were really taking place. (p. 354)

Significantly, this observation is closely connected to the chaplain's definition of the phenomenon of *jamais vu*, a process where unfamiliar events acquire a feeling of familiarity (see extract 15). This seems to refer to the process where illusory, nonfactual and counterfactual domains seem to take over the actual, the reality, in such a way that what is unexpected and unusual becomes the generally accepted rule. In linguistic terms, we can define it as a process whereby what is usually expressed in marked terms by means of negative clauses and counterfactuals, becomes the unmarked more usual option.

5.3.2.2. *Negative accommodation*

By accommodation, Werth (1995c, p. 404) describes a phenomenon where new information is presented in an unconventional way, that is, by means of subordinate clauses or other elements that are normally the vehicle for background information.[26] Werth (p. 404) observes that in these cases, "the so-called presuppositional content does not reflect backgrounded information, and therefore has to be regarded as assertive." Negation can be a vehicle for unconventional assertion when it does not carry out the function of rechanneling information, as described in the previous sections, but when it is used to introduce an item that is simultaneously denied. As I argued previously, in this case, negation performs an identifying function—and not an individuating function in descriptions. The identification of accommodation in a long stretch of discourse, such as a novel, is not always easy, as there are cases in which it is not possible to establish a clear cut distinction between old and new information according to grammatical units such as clauses. There are cases, however, that seem to illustrate clearly what Werth has described as negative accommodation, understood as a function of negation where an item was simultaneously introduced in the discourse and denied. The following is a significant example (see also the discussion in Sections 3.3.4. and 3.4.2. in Chapter 3). Extract (23) is part of a longer description of how people *do not die* in the hospital, thus describing how they die outside the hospital, in the war or in other violent situations.

(23) There was none of that crude, ugly ostentation about dying that was so common outside the hospital. They did not blow up in mid-air like Kraft or the dead man in Yossarian's tent, or freeze to death in the blazing summertime the way Snowden had frozen to death after spilling his secret to Yossarian in the back of the plane
"I'm cold," Snowden had whimpered. "I'm cold."
"There, there," Yossarian had tried to comfort him. "There, there."
They didn't take it on the lam weirdly inside a cloud the way Clevinger had done. They didn't explode into blood and clotted matter. They didn't drown or get struck by lightning, mangled by machinery or crushed in landslides. They didn't get shot to death in hold-ups, strangled to death in rapes, stabbed to death in saloons, bludgeoned to death with axes by parents or children or die summarily by some other act of God. Nobody choked to death. People bled to death like gentlemen in an operating room or expired without comment in an oxygen tent. There was none of that tricky now-you-see-me now-you-don't business so much in vogue outside the hospital, none of that now-I-am-and-now-I-ain't. (pp. 212–213)

The negative clauses in this extract clearly carry out the function of introducing an item (how people die, for example, by blowing up or freezing to death) and denying its truth within the domain of the hospital. The denial of violent deaths in itself, however, evokes a second set of domains where these deaths would be the case, that is, outside the hospital, at war.[27] This world outside the hospital includes, not only war, but also other violent situations, which are described in detail so as to provide a complete list of possible violent deaths people face every day. In this sense, the negative propositions evoke frame knowledge associated with both domains, that is, expectations about (not violent) death in a hospital and expectations about (violent) death outside a hospital.

The extract is significant because one of the themes of the novel, death, is not dealt with directly or explicitly, but, rather, it is mentioned indirectly, by a strategy of exclusion. In this strategy, the function of negation as introducing and denying an item at the same time plays a crucial role. A similar argument is followed in the analysis of a relevant example in Chapter 3, where the German enemy is described in negative terms, in the same way as death in war is described in extract (22).

As a reader, one may wonder why significant themes as death, war, and the definition and description of the enemy are not dealt with by means of straightforward linguistic descriptions, but rather, by means of an indirect presentation by negating their applicability in different situations. It can be argued that this strategy is one more within a general procedure by which a tragic subject is dealt with by indirect means. It is precisely this indirectness, which places a heavy load of interpretation on the reader, that makes the tragic aspect of the novel seem greater once it has become clear. I argue that the indirect presentation of a tragic content is also a strategy observable in the recurrent use of contradiction, which is discussed in the following sections.

5.4. CONTRADICTION: NEGATION AS SUBWORLD THAT BLOCKS THE FLOW OF INFORMATION

It has already been pointed out that contradiction can be of two types in *Catch-22*, depending on whether it contributes to the function of rechanneling information or whether it seems to block the flow of information. This distinction has to do with the possibility of interpreting the contradiction as meaningful at a low processing level, that is, without having to evoke higher level knowledge frames that form part of the novel as a whole. Thus, the types of contradictions that belong to the former type lead to a resolution of the apparently contradictory propositions in either of two ways. In the first type of structure, one of the contradictory terms is finally favored over the other, thus illustrating the account of contradiction defended by Sperber and Wilson (1986, p. 339). In the second type of structure, the contradiction is interpreted as meaningful by determining the validity of the two contradictory terms in two different domains (see Escandell, 1990; Fauconnier, 1985). The latter type of contradiction cannot be interpreted as meaningful in these ways; in a pragmatic sense, however, it can be argued that these apparently irresolvable contradictions *convey more meaning than is literally said*, but this meaning can only be recovered by interpreting the contradictions within the discourse of the whole novel.

Strictly speaking, I do not deal with logical contradiction, that is, a compound proposition where one of the conjoins contradicts the other, but with what can be called *discourse contradiction*. By this is meant a function of a negative clause where a speaker denies the truth of a proposition that has previously been asserted as true. This obviously reveals some kind of inconsistency, which is manifested explicitly or inferred from the context of utterance.

5.4.1. Contradiction that Contributes to the Rechanneling of Information in Discourse

The following extracts illustrate the types of contradictions that rechannel information in discourse. Extract (24) is part of a conversation between Orr and Yossarian that shows Orr's peculiar way of understanding conversational rules and maxims. The passage is long, but it is crucial for the understanding of key aspects of the novel.

(24) "I wanted apple cheeks," Orr repeated. "Even when I was a kid I wanted apple cheeks someday, and I decided to work at it until I got them, and by God, I did work at it until I got them, and that's how I did it, with crab apples in my cheeks all day long." He giggled again. "One in each cheek."
"Why did you want apple cheeks?"
"I didn't want apple cheeks," Orr said. "I wanted big cheeks. I didn't care about the color so much, but I wanted them big. I worked at it just like one of those crazy guys you read about who go around squeezing rubber balls all day long just to strengthen their hands. In fact, I was one of those crazy guys. I used to walk around all day with rubber balls in my hands, too."
"Why?"
"Why what?"
"Why did you walk around all day with rubber balls in your hands?"
"Because rubber balls—" said Orr.
"—are better than crab apples?"
Orr sniggered as he shook his head. "I did it to protect my good reputation in case anyone ever caught me walking around with crab apples in my cheeks. With rubber balls in my hands I could deny there were crab apples in my cheeks. Every time someone asked me why I was walking around with crab apples in my cheeks, I'd just open my hands and show them it was rubber balls I was walking around with, not crab apples, and that they were in my hands, not my cheeks. It was a good story. But I never knew if it got across or not, since it's pretty tough to make people understand you when you're talking to them with two crab apples in your cheeks." (pp. 34–36)

This conversation, of which the present extract is only a part, is a good example of what I define as *discourse contradiction*. Orr first tells Yossarian he wants apple cheeks, an assertion he later denies or contradicts by saying it is not really apple cheeks he wants, but big cheeks (*I wanted apple cheeks* and *I didn't want apple cheeks*). The relation between the two clauses can be described as contradictory because the second clause inverts the truth value of the first one. The contradiction is resolvable in the terms proposed by Sperber and Wilson (1986) in that the second term (*I didn't want apple cheeks*) is ultimately favored over the first one; this is confirmed by an explanation (*I wanted big cheeks*) that justifies the reformulation.

The contradiction is complemented by a pattern where a deceitful appearance is presented that distorts the perception of a given situation, thus following a similar procedure to the ones described in previous episodes. In this case, Orr foregrounds a fact (carrying rubber balls in his hands) in order to conceal another one (carrying crab apples in his cheeks). Neither Yossarian nor the reader can infer the implied meaning of this image till the end of the novel, when Yossarian is told that Orr has reached Sweden by rowing a small boat after he has disappeared in a plane crash. This knowledge enables the reader to understand retrospectively the meaning of Orr's cryptic conversations, in which it is revealed that he is trying to say that by doing something (carrying rubber balls in his hands; crashing his plane) he is actually aiming at something else (trying to have big cheeks; escaping to Sweden).

It makes sense to say that the contradiction mentioned previously should be understood in the light of this argument. The process of interpretation shows, as in previous examples, that the processing of negation cannot be limited to the negative clause itself, but that it is crucial to consider the function of the negative term within longer stretches of discourse.

A different example can be observed in extract (25), where Milo, the mess officer, is trying to find a way of getting rid of a harvest of cotton he has bought. Yossarian suggests he should ask the government for help.

> (25) "It's a matter of principle," he explained firmly. "The government has no business in business, and I would be the last person in the world to ever try to involve the government in a business of mine. But the business of government is business," he remembered alertly, and continued with elation. (p. 337)

In his argument, Milo changes from asserting that the government has no business in business to asserting the contrary, that the government's business is business. The difference here lies in the domain in which each of the two assertions is applicable; in the first case, it is understood that the government should not take part in business matters; in the second case, that the government should interfere in order to solve problems like the one Milo is facing. Seen in terms of the development of the discourse, however, the second term of the contradiction (*But the business of government is business*) is the term that is ultimately favored by the speaker, again illustrating the inferencing process argued by Sperber and Wilson. A point to be noticed in this example is that the term that is finally favored is, in fact, the positive term, and not the negative. This illustrates the fact that was commented on in Chapter 2 that both positive and negative utterances can perform a function of contradiction.

Some contradictions not only lead to the eventual favoring of one of the terms of the contradiction, but at the same time, it is the case that each term of the contradiction is understood to be applicable in a different cognitive domain (spatial, temporal, epistemic, or other).

In extract (26), Major Major is trying to find something to tell Yossarian about the increase in the number of missions.

> (26) What could you possibly say to him? Major Major wondered forlornly. One thing he could not say was that there was nothing he could do. To say there was nothing he could do would suggest he would do something if he could and imply the existence of an error of injustice in Colonel Korn's policy. Colonel Korn had been most explicit about that. He must never say there was nothing he could do. "I'm sorry," he said. "But there's nothing I can do." (p. 135)

Major Major's words to Yossarian contradict his thoughts (*He must never say there was nothing he could do; he said "But there's nothing I can do."*). This leads to a reformulation of what was assumed to be the conclusion of his thought process. The contradiction can be interpreted as a contradiction between internal worlds of the character, an obligation world (what he has been instructed to say), and an epistemic world (what he knows he can and cannot do). As in the previous examples, the second term is favored over the first one. It is interesting to observe that Major Major is aware of the implications of using a negative, as negating something presupposes the corresponding affirmative: *To say there was nothing he could do would suggest he would do something if he could.*

In other cases, however, a resolution does not seem to be possible in such immediate terms, and the problem of the applicability of two terms in different domains becomes irreconcilable.

> (27) The only one with any right to remove his belongings from Yossarian's tent, it seemed to Major Major, was Yossarian himself, and Yossarian, it seemed to Major Major, had no right. (p. 132)

This extract illustrates a contradiction between the propositions *Yossarian was the only one with any right to remove his belongings from Yossarian's tent* and *but Yossarian had no right.* Each of the terms is valid in a different cognitive domain. Legally, Yossarian has no right to remove another soldier's belongings from his tent because he lacks the authority to do so; however, on a more personal level, since the dead soldier's belongings are in Yossarian's tent, it makes sense to assume that Yossarian should be entitled to dispose of the things as best he thought. A solution to the dilemma is not provided.

The examples discussed so far illustrate contradictions that perform a discourse function that leads to a reformulation of a previous proposition, with the general effect of rechanneling the information, or contradictions that require the interpretation that the two contradictory terms are applicable in two different cognitive domains.

5.4.2. Contradiction That Does Not Contribute to the Rechanneling of Information: Circular Logic as a Communicative Short Circuit

Unlike the contradictions discussed in the previous section, the following passages illustrate contradictions that require the acceptance of the contradictory nature of the structure, with no possibility of interpreting an applicability of the terms in different domains. The frequency of this type of contradictory structure is not necessarily high; however, it is crucial for the understanding of the novel, since it occurs in connection with the theme that provides the title to the novel, catch-22. The catch is defined by means of a circular argument; extract (28) is part of a conversation between Doc Daneeka and Yossarian where Yossarian is inquiring about the possibilities of being grounded and being sent back home. Doc Daneeka answers he can be grounded if he is crazy, but there is a catch.

(28) There was only one catch and that was Catch-22, which specified that a concern for one's own safety in the face of dangers that were real and immediate was the process of a rational mind. Orr was crazy and could be grounded. All he had to do was ask; and as soon as he did, he would no longer be crazy and would have to fly more missions. Orr would be crazy to fly more missions and sane if he didn't, but if he was sane he had to fly them. If he flew them he was crazy and didn't have to; but if he didn't want to he was sane and had to. Yossarian was moved very deeply by the absolute simplicity of this clause of Catch-22 and let out a respectful whistle. (pp. 62–63)

Catch-22 has been already described in the introduction, but comment is made here on it briefly once more by focusing on how it is described in this passage. The description incurs in contradiction.

a.   Orr would be crazy to fly more missions and sane if he did not.
b.   But if he was sane he had to fly them.
c.   If he flew them he was crazy and did not have to.
d.   But if he did not want to he was sane and had to.

As explained in the introduction, the catch-22 can be summarized as a circular logical process shown as follows.

a.   If you are crazy you can be grounded.   If a then b.
b.   If you want to be grounded you have to apply.   If b then c.
c.   If you apply you are not crazy.   If c then d (= not a).

The process involves a contradiction between the first premise *a* (*if you are crazy*) and the conclusion of the argument *d = not a* (*you are not crazy*); this makes it impossible for the proposition *you can be grounded* ever to be applicable.

A similar reasoning pattern seems to be adopted systematically by Luciana, the Italian girl Yossarian meets one night and with whom he falls in love. Yossarian asks her to marry him, and the conversation develops as shown in extract (29).

(29) "Tu sei pazzo," she told him with a pleasant laugh.
     "Why am I crazy?" he asked.
     "Perchè non posso sposare."
     "Why can't you get married?"
     "Because I am not a virgin," she answered.
     "What has that got to do with it?"
     "Who will marry me? No one wants a girl who is not a virgin."
     "I will. I'll marry you."
     "Ma non posso sposarti."
     "Why can't you marry me?"
     "Perchè sei pazzo. "

"Why am I crazy?"
"Perchè vuoi sposarmi."
Yossarian wrinkled his forehead with quizzical amusement. "You won't marry me because I'm crazy, and you say I'm crazy because I want to marry you? Is that right?"
"Si"
"Tu sei pazz'!" " he told her loudly. (pp. 205–206)

As in the example from the catch-22, Luciana's argument is circular and, consequently, leads nowhere beyond the internal self-referentiality of the system, as each of its parts leads circularly to the next. Nash (1985, p. 111) points out that this type of reasoning is typical of certain psychiatric patients.

The catch-22 can be said to represent symbolically the closed system of the world of *Catch-22* itself. The soldiers are caught in a trap that prevents them from leaving the island and going back home. The reader is told that Colonel Cathcart has successively increased the number of missions the men have to fly as a way of displaying power. When he decides to send the men back, it is too late, since sending the whole squadron back and asking for replacements would be suspicious. As a consequence, the men are doomed to a certain death, and little by little Yossarian sees how his friends are killed or disappear. The only escape is to opt out of the system, as Orr does by rowing to Sweden, or as Yossarian does at the end of the book, by deserting and running away.

Thus, the understanding and processing of the contradiction of the catch-22 lies in its implications as an instrument of abuse in the hands of the military officers: it gathers up the series of conflicts discussed in different episodes, especially the themes of arbitrariness and of pretense. Indeed, the catch-22 does not exist as part of the military regulations, but it is used as an excuse to carry out actions on an arbitrary basis. The catch-22 describes a system, and as such, it goes beyond the structure of power of the higher officers and illustrates the workings of a complex society.

5.5. LEXICAL OPPOSITION AND CONTRARIETY

An analysis of opposition by means of lexical negation in *Catch-22* reveals that this phenomenon is closely connected to the phenomenon of contradiction described in the previous section. Thus, in this section, extracts that present examples of lexical opposition of the types described in Section 2.3.4. in Chapter 2 are commented on. Some of the examples present a combination of syntactic and lexical negation. For practical reasons, I use the general term *contradiction* to refer to relations between lexical opposites as well as to refer to contradictory propositions such as those discussed in the previous section.

I argue that the relations of lexical opposition discussed in this section can be interpreted as meaningful by means of identifying the domain within which each of the opposite terms can be said to apply, following the same procedure described for certain types of contradictions.[28] For this purpose, a schema theoretic approach to the description of lexical opposition is adopted. As already done in Chapter 4, it is suggested that each term in an opposition can be understood to evoke a frame or schema. This approach can be incorporated into Werth's (1995c/1999) text world model to provide a detailed and systematic account of how frame knowledge is processed in discourse. For reasons of space, we only discuss a selection of examples that illustrate some of the relevant types of opposition that can be encountered in the novel. Each example makes reference to a particular topic or subtopic of the novel, such as madness, power, personality, business deals, war, and so on.

Relations of contrariety are a vehicle for parody and criticism of attitudes and behavior in society. Extract (30) is about racism.

(30) "Some of those invitations were mighty generous, but we couldn't accept any because we were Indians and all the best hotels that were inviting us wouldn't accept Indians as guests. Racial prejudice is a terrible thing, Yossarian. It really is. It's a terrible thing to treat a decent, loyal Indian like a nigger, kike, wop or spic." Chief White Halfoat nodded slowly with conviction. (p. 60)

In this extract there is a double opposition, the first being the contradiction between the hotels inviting Indians and their refusal to accept them because they are Indians, and the second, the contradiction between Chief White Halfoat's criticism of racism and his racist attitude towards other races.

Extract (31) is a contradiction that focuses on the petty struggles for power on the part of the higher officers. For General Peckem, the enemy are not really the Germans but, rather, another General on the island, General Dreedle. General Peckem spends his time planning "offensives" against General Dreedle in order to have all the power in his hands.

(31) "Yes, I know I understand. Our first job is to capture Dreedle away from the enemy. Right?" General Peckem laughed benignly, "No, Scheisskopf. Dreedle's on our side, and Dreedle is the enemy." (pp. 408–409)

The contradiction in (31) confirms Yossarian's suspicions that the enemy is anybody who wants to kill him and the rest of the soldiers, no matter what side they are on. Indirectly, this seems to suggest that the higher officers are on an equal status with the Nazis.

Contradiction reveals conflicts regarding some characters' incapacity to distinguish between moral values. An example is extract (32), about Hungry Joe.

(32) Every time Colonel Cathcart increased the number of missions and returned
Hungry Joe to combat duty, the nightmares stopped and Hungry Joe settled down
into a normal state of terror with a smile of relief. Yossarian read Hungry Joe's
shrunken face like a headline. It was good when Hungry Joe looked bad and
terrible when Hungry Joe looked good. Hungry Joe's inverted set of responses
was a curious phenomenon to everyone but Hungry Joe, who denied the whole
thing stubbornly. (p. 73)

Hungry Joe is described as having an inverted set of responses, in such a way that
when the number of missions is increased he looks good, whereas when he is not
on combat duty, he looks awful. This leads to an inversion of the equations that
good = good and bad = bad, which are substituted by good = bad, bad = good. A
similar process can be said to take place in Captain Flume, who is unable to
distinguish between his dreams and his waking states. Captain Flume tries hard to
stay awake all night because Chief White Halfoat has threatened to kill him during
the night.

(33) Each night after that, Captain Flume forced himself to keep awake as long as
possible. He was aided immeasurably by Hungry Joe's nightmares. Listening
so intently to Hungry Joe's maniacal howling night after night, Captain Flume
grew to hate him and began wishing that Chief White Halfoat would tiptoe up
to his cot one night and slit his throat open for him from ear to ear.
Actually, Captain Flume slept like a log most nights and merely dreamed he
was awake. So convincing were these dreams of lying awake that he woke from
them each morning in complete exhaustion and fell right back to sleep. (pp.
76–77)

In this extract, the distinction between dreaming/being asleep and being awake
becomes blurred by means of a recurring pattern of dreams within dreams, where
the captain dreams he is awake and never gets enough rest.
The following are further examples of contradictions involving lexical opposites.

(34) "Oh, shut up," Dunbar told Clevinger. Dunbar liked Clevinger because Clevinger annoyed him and made the time go slow. (p. 29)

(35) Dunbar loved shooting skeet because he hated every minute of it and the time
passed so slowly. "I think you're crazy," was the way Clevinger had responded
to Dunbar's discovery. (p. 52)

(36) Ordinarily, Yossarian's pilot was McWatt, who, shaving in loud red, clean
pajamas outside his tent each morning was one of the odd, ironic, incomprehensible things surrounding Yossarian. McWatt was the craziest combat man of them
all probably, because he was perfectly sane and still did not mind the war. (p.
80)

(37) He woke up blinking with a slight pain in his head and opened his eyes upon a world boiling in chaos in which everything was in proper order. (p. 186)

(38) Colonel Cathcart was not superstitious, but he did believe in omens. (p. 267)

(39) You see, Italy is really a very poor and weak country, and that's what makes us so strong. (p. 309)

(40) This sordid, vulturous, diabolical old man reminded Nately of his father because the two were nothing at all alike. (p. 311)

(41) The chaplain was sincerely a very helpful person who was never able to help anyone. (p. 346)

(42) "My only fault," he observed with practiced good humor, watching for the effect of his words, "is that I have no faults." (p. 405)

The opposites presented in the previous extracts are collected in Table 5.7 and classified according to the type of relation that holds between the opposite terms.[29]

Each of the contradictions can be resolved at a higher level of interpretation by determining the domain within which each of the terms is applicable. As argued in Chapter 4, we can understand the higher level schematic structure as a meta-MOP (Memory Organization Packet) in Schank's terms. This higher level structure would collect information about lower level schematic packets regarding societal, personal, and situational information. Inconsistencies between the different domains can lead to contradictions of the types that have been discussed so far. We will not go into details of the description here, but the reader is referred to Chapter 4 for a discussion of how this type of analysis can be carried out.

TABLE 5.7.

Opposites	Type of Opposition
Like/Annoy	Weak opposites or contraries
Love/Hate	Strong or polar opposites
Crazy/Sane	Strong or polar opposites
Chaos/Order	Strong or polar opposites
Not superstitious/Believes in omens	Weak opposites or contraries
Weak/Strong	Strong or polar opposites
Remind/Different	Weak opposites or contraries
Helpful/Not helpful	Contradictories
Have faults/Not have faults	Contradictories

Other contradictory relations are created by means of what Fillmore (1985, p. 243) calls across-frame negation, or opposition by means of contrasting apparently disparate terms. According to the classification followed previously, the type of opposition between these terms is that of contrariety or weak opposition. Below are two examples.

(43) Kraft was a skinny, harmless kid from Pennsylvania who wanted only to be liked, and was destined to be disappointed in even so humble and degrading an ambition. Instead of being liked, he was dead. (p. 74)

(44) In short, Clevinger was one of those people with lots of intelligence and no brains, and everyone knew it except those who soon found it out. (pp. 90–91)

The two extracts present unusual oppositions between terms that would not normally be related as opposites: in extract (43), we have the opposition between be liked/be dead, and in (44), be intelligent/have no brains. The unusual contrasts make the extracts slightly humorous.

Similarly, contradiction is used as a means of power abuse on the part of higher officers. Below is part of an extract where some officers are interrogating Nately's whore. They do so by obliging her to say *uncle*.

(45) "You still don't understand, do you? We can't really make you say uncle unless you don't want to say uncle. Don't you see? Don't say uncle when I tell you to say uncle. Okay? Say uncle."
"Uncle," she said.
"No, don't say uncle. Say uncle."
She didn't say uncle.
"That's good!" (p. 445)

This apparently nonsensical exchange reveals the sadistic side of higher officers, as they seem to enjoy trying to make somebody confess to something when that person does not want to confess. The girl does not understand, and does not mind "confessing," or saying "uncle," which obviously annoys the officers, who do not enjoy the interrogation.

The pervasive character of inverted values in society is revealed toward the end of the novel, where the narrator clearly questions the reliability of so many apparent values we take for granted, without questioning the sincerity of the act behind them. This is summarized in extract (46).

(46) How many winners were losers, successes failures, rich men poor men? How many wise guys were stupid? How many happy endings were unhappy endings? How many honest men were liars, brave men cowards, loyal men traitors, how many sainted men were corrupt, how many people in positions of trust had sold their souls to blackguards for petty cash, how many had never had souls? How

many straight-and-narrow paths were crooked paths? How many best families were worst families and how many good people were bad people? (pp. 520–521)

What has been suggested, more or less implicitly, earlier in the novel is now questioned directly, thus revealing a kind of world where nobody and nothing can be trusted, since anything we may have valued positively may turn out to be the opposite. This has been shown to be the case for many situations in the novel, which are discussed under various headings in the present chapter. The process by which values are inverted is described as being discovered by the chaplain in the following way.

(47) The chaplain had sinned, and it was good. Common sense told him that telling lies and defecting from duty were sins. On the other hand, everyone knew that sin was evil, and that no good could come from evil. But he did feel good; he felt positively marvelous. Consequently, it followed logically that telling lies and defecting from duty could not be sins. The chaplain had mastered, in a moment of divine intuition, the handy technique of protective rationalization, and he was exhilarated by his discovery. It was miraculous. It was almost no trick at all, he saw, to turn vice into virtue and slander into truth, impotence into abstinence, arrogance into humility, plunder into philanthropy, thievery into honor, blasphemy into wisdom, brutality into patriotism and sadism into justice. Anybody could do it—it required no brains at all. It merely required no character. (p. 459)

To sum up the observations made in the preceding sections, contradiction is a recurrent discursive strategy in *Catch-22*, and it can be carried out either by means of syntactic negation or by means of lexical opposition. In both cases, the effect in discourse is that of creating a block in the communicative flow that has the form of an apparently unresolvable information structure. I suggest that the contradictions can be resolved at a higher level of interpretation because each of the terms can be interpreted as meaningful in a different domain, or because the contradiction in itself leads to a questioning of procedures we can observe in society. In both cases, contradiction can trigger schema refreshment by leading to a questioning of taken-for-granted assumptions about how things are in the world.

5.6. CONCLUSION: THE ROLE OF NEGATION IN THE CREATION OF A WORLD VIEW IN JOSEPH HELLER'S NOVEL *CATCH-22*

From the discussion of this chapter, several concluding remarks can be made. With regard to the theoretical model proposed for the analysis of negation, a text world approach to negation makes it possible to consider the functions of negation in discourse by focusing on its cognitive properties, an aspect that tends to be overlooked in other discourse-based accounts of negation. The discussion in the present chapter shows that it is crucial to consider the cognitive properties of negative statements to understand the kind of worldview that is being projected by a given discourse. Furthermore, a text world account of negation also focuses on the relations between negative statements and preceding discourse, which is responsible for creating the assumptions that are defeated by means of negation. This means that the analysis of negation in discourse cannot be restricted to its semantic and structural analysis, but, rather, it requires a view of negation as a complex and dynamic discourse process.

Considering the role of negation in the novel *Catch-22* as a whole, by means of an analysis of examples such as the ones discussed throughout this book, it can be argued that negation plays a crucial role in the expression of an underlying conflict in the fictional world. This conflict is manifested in two different ways. The first is negation in the function of rechanneling information and the second is negation involved in contradictory structures. In the case of negation that rechannels information, the underlying conflict is manifested as a lack of fit between the status quo in the fictional world and the perception of such facts by characters or the representation of facts in illusory domains; the conflict is deepened by the fact that the representations in the illusory domains are taken as if they were real. Thus, I argue that negation, in some cases, denies the presence or existence of entities that had previously been taken for granted in the fictional world, such as, for example, the catch-22 itself. In other cases, negation contributes to the projection of hypothetical, modal or future subworlds which, in spite of their nonfactual character, seem to presuppose that the nonfact has actually taken place. Negation also modifies cognitive subworlds so as to express some characters' inability to process real facts as real. In linguistic terms, these examples seem to show that negation foregrounds nonevents and nonstates in cases where this goes against commonly held expectations derived from knowledge of the world, information that has been provided previously in the novel or cultural knowledge. Furthermore, the recursive denial of world building and descriptive information that had been previously set up in the text world, reveals a pattern in which the fictional world seems to suffer a process of deconstruction, that is, a process where the parameters that had been accepted as valid and true are progressively denied such validity. In this sense, the world of *Catch-22* comes close to the type of postmodern fictional world that Doležel defines as one that lacks authentication because it infringes the law of non-contradiction (see Doležel, 1989). Strictly speaking, the fictional world of *Catch-22* is not inconsistent in this sense. The breakdown is not in the reality of the fictional world but in the language system that describes the fictional world.

With regard to the denial of narrative function-advancing propositions, negatives are used to invert the logic of cause–effect relations in such a way that standard assumptions that may be held about this type of logical relation are not applicable in the fictional world of *Catch-22*. In more general terms, negatives deny function-advancing material in order to defeat expectations that have arisen with regard to the development of events, such as, for example, how a certain mission is going to take place.

In the denial of descriptive material, I observe the consistent and systematic recursion of situations that foreground the nonstate. I argue that these situations could be said to represent, in some cases, a conflict between an illusory appearance set up by means of deceit and the status quo in the fictional world. In terms of Ryan's (1991b) model of conflict in fictional worlds, it reflects a conflict between characters' wish worlds to be perceived by others in a certain way and the reality behind the appearance. I also argue that there seems to be a progression in the novel in such a way that the appearance, no matter whether true or false, progressively takes over reality, or becomes a reality, as it is interpreted as such by the characters.[30] This phenomenon is summarized by means of two processes where *what is* is substituted by *what is not*, and *what cannot be* becomes *what is.*

In other cases, the conflict expressed by means of negatives reflects, rather, an opposition between the wishes of different characters, typically, between the higher officers' interests and those of the common soldiers, who want to go home once they have completed their number of missions. The possibility of leaving combat duty proves to be impossible, due to the existence of a *catch-22*, a circular logic trap that prevents soldiers from being grounded. This leads to the second type of negative function discussed, contradiction.[31] Whereas negatives of the first type are used to present events and properties that are not real *as if* they were real, in the case of contradiction, the distinction between opposites is blurred, and standard assumptions about logical relations that do not infringe the law of non-contradiction are not respected.[32] This reveals a further type of conflict in the fictional world, which has to do with an internal lack of consistency exemplified by the logic of the catch-22.

I also argue that the function of negation described in these terms contributes to the development of a pattern of defamiliarization that leads to a questioning of taken-for-granted assumptions about the way in which things are done in well-known areas of society, such as the military, religion, law and justice, business, economy, and war, to mention the more prominent examples. Such a questioning can be said to be a potential trigger for schema refreshment, although it is also pointed out that schema refreshment is reader and culture dependent.

Turning now to the claim made in the introduction of this book, that negation contributes to the creation of a worldview in the novel *Catch-22*, it can be said that the recursive presence of negatives that foreground nonevents and nonstates against commonly held assumptions, and the recursion of contradiction have two peculiar effects. The first effect is that of making nonevents and nonstates seem more salient than events and states, contrary to what is the general tendency in the cognitive perception and organization of experience; the second is to blur the distinction between opposite terms and to deconstruct the internal coherence of logical structures by means of contradiction, thus generating instability in the way characters perceive reality and interact with it. The worldview that emerges from a reading of the novel based on the observations made is one that challenges assumptions about wars and the role of institutions, and leaves little hope for the freedom of action of individuals, who seem to be caught half way between an inability to understand and react to reality on the one hand, and the (il)logic of institutionalized behavior illustrated by the catch-22, on the other. It is also a world view that questions the validity of assumptions regarding the relation between language as a system of organizing and interpreting reality and reality itself. This point is particularly significant because of the privileged status of the language, the discourse, over the reality.

The world of *Catch-22*, both in its tendency to accept the nonfactual as factual and in its recurring contradictions, seems to have developed the features of a closed system that has rules of its own, independent from reality as we know it. Indeed, several critics have observed that the world of *Catch-22* displays the characteristics that define some of the major images of the novel[33]: the circularity of the catch-22 and the closed system (also circular) of the soldier in white, who is kept alive by means of two glass jars connected to different parts of his body and which are exchanged as soon as one of them is full. As a closed system, the world of *Catch-22* is a system in a partial state of entropy (see Tucker, 1984). It can be argued that the recurrent use of negation in the discourse of the novel contributes to the foregrounding of the ontological characteristics of a fictional world that tends to *termic death*, to *total entropy*. In this sense, the process whereby what is is replaced by what is not, and the process whereby the distinction between opposites becomes blurred and logical processes break down, are signs of a self-reflective, closed system with its own rules that is not modified by external influences. Thus, none of the soldiers leave the island alive, and none of the scandalous events carried out in the squadron can be modified by means of action from an external agent. This leads to the conclusion that the world of *Catch-22* seems to have no escape other than opting out of the system altogether, for instance, by means of desertion, as Orr and Yossarian do in different ways.

NOTES

1. See Sections under 2.6. in Chapter 2.
2. See Chapter 4.
3. See the second part of Chapter 3 for a discussion of Ryan's model of narrative discourse.
4. See Chapter 6 for a discussion of the classification into negation types.
5. For more details regarding the types of negation in the corpus, see the appendix.
6. See the discussion of the examples in the last sections of Chapter 2 as illustrations of this view of negation in the discourse of *Catch-22*.
7. See Chapter 4 for a discussion of how humor can be distinguished from nonsense because of the possibility of a higher level resolution of incongruity, which in nonsense remains unresolved.

8.   By marked negation, I mean negative utterances that are grammatical but sound odd because they deny commonly held assumptions about how things are in this world.

9.   This was pointed out to me by Geoffrey Turner, to whom I am very grateful.

10.   For a definition of syntactic negation and lexical negation, and for a discussion of different classifications of categories of negative words, see Sections under 2.3. in Chapter 2.

11.   For an account of the possible relations that can be observed between negative and affirmative clauses in discourse, see Jordan (1998).

12.   For a definition of *common ground* see the discussion of Werth's model in Chapter 3.

13.   For an account of the uninformativeness of negative clauses, see Leech's (1983, pp. 100–102) *principle of negative uninformativeness* and *sub-maxim of negative uninformativeness*. The former describes the general uninformative character of negative propositions as compared to their positive counterparts. The latter establishes that "a negative sentence will be avoided if a positive one can be used in its place" (Leech, 1983, pp. 100–102). Horn (1989, p. 201) further develops an approach to negative uninformativeness based on the assumption that the Gricean maxims of quantity and relation are flouted so as to create an implicature that has become conventionalized for all negative utterances. This implicature is precisely the uninformative character of the negative with respect to the affirmative. Thus, Horn (1989, p. 201) observes that

> Negative propositions are typically, but not necessarily, less specific and less informative than positive propositions. . . . However, the real asymmetry is located, not in the relation to positive propositions, but in the relation of speaker denials to assertions.

See also Kuno (1992a, p. 14) and (1992b, pp. 4–6) for the *ban on questions that solicit uninformative answers*, and Givón (1993, pp. 191–193).

14.   See Givón (1993, p. 197) and the discussion in Section 2.3.3. in Chapter 2 of the present book.

15.   This was pointed out to me by Chris Butler (personal communication), to whom I am grateful.

16.   This is one of the most significant scenes in the novel, and is loaded with deliberate symbolism. Yossarian has refused to put on his uniform for the funeral, as it was stained with the blood of the dead bombardier. The death of this soldier has made Yossarian aware of what death (and life) is, and he sits on the tree, naked, watching the ceremony. Milo approaches him in order to talk to him, and Yossarian says it's the tree of life he's sitting on. Milo, with his innate pragmatism, rejects this view, and says it is not the tree of life but a chestnut tree.

17.   In Chapter 2, the distinction between contraries is discussed. An example is black/white and contradictories black/not black. The tendency to categorize in terms of contradictoriness, rather than contrariety, is pointed out.

18.   See Section 4.2.5. in Chapter 4 for a discussion of Schank's (1982) model.

19.   See Apter's (1982) observation that in order for humor to take place, it is necessary to have a clash between an illusory and a factual domain, what he calls *the appearance* and *the reality*. It is also necessary that the appearance seems to be *more than* the reality.

20. See the discussion of Schank's (1982) model and the application to selected extracts in Chapter 4.

21. This view of negation fits in with Givón's notion of the presuppositional nature of negation in discourse.

22. See Semino's (1994, 1997) application of Schank's model to poetic worlds.

23. This was pointed out to me by JoAnne Neff (personal communication), to whom I am grateful.

24. See also the episode where Luciana, the girl Yossarian picks up one night in Rome, asks him to ask her to write her telephone number for him. When he does so, she replies angrily "Why, so you can tear it up into little pieces as soon as I leave?" Her attitude seems to form part of what I define as *negative determinism*, in that she projects the possibility of a nonfactual state of affairs becoming actual by introducing it into the conversation, when it was completely unnecessary. What is presented as crazy and absurd actually takes place. Thus, Yossarian tears up her telephone number into little pieces, something he later regrets, but it is too late.

25. The episode brings to mind the long-standing philosophical argument on the status of objects that are not perceived by external observers, from Berkeley to Heisenberg.

26. See Section 3.3.4. in Chapter 3 for a detailed description of the phenomenon.

27. See Section 2.2.3. in Chapter 2 for a discussion of how negation is understood to affirm in psychological experiments.

28. For a discussion of this view from a schema theoretic perspective, see Chapter 4.

29. See Chapter 2 for a classification of lexical opposites.

30. See Greenberg (1966) and Tucker (1984) for a discussion of how the vacuity or emptiness that characterizes the world of *Catch-22* can be explained in terms of entropy and information theory. According to these authors, the progression of the story in *Catch-22* is a process of disintegration, which leads to inertia, and, metaphorically, to "termic death." Hence, the recurrent use of images that have to do with cold, rain, and snow; when the soldiers lose their hope of ever leaving the island, we are told that the weather changes and there are no more beautiful days. Snowden is the character who symbolically represents human vulnerability and mortality, and his death haunts Yossarian throughout the novel.

31. See Aguirre (1991) for a discussion of how the violation of the law of the excluded middle is a characteristic of postmodern fiction. Without being strictly postmodern, *Catch-22* presents a world where the entities that inhabit it can be defined in contradictory terms. The novel is not truly postmodern because the contradictions are resolvable by inferencing.

32. The blurring between opposites is particularly significant in the case of certain oppositions, such as alive/dead and crazy/sane. The contradictions on madness lead to a reflection on the difficulties in defining the boundaries between sanity and craziness. See Rosenhan (1973) for a description of an experiment carried out by a team of psychologists who managed to be registered as psychiatric patients in several mental institutions, but had great difficulties in proving that they were sane and being released. In linguistics, it has traditionally been observed that opposites in some way "attract each other." See, for instance, Cruse (1986, p. 197).

33. See, for example, Tucker (1984) for an account of *Catch-22* in terms of information theory and the notion of *entropy*. For a discussion of the notion of entropy in discourse and text, see Bernárdez (1995, p. 121).

6

Conclusions

The non-event is pragmatically—and indeed grammatically—the oddest. This must be so because if an event did not occur *at all*, why should one bother to talk about a specific individual who "participated" in that non-event?

Givón (1993, p. 191)

It has been said more than once that in a text negation blocks the flow of information....

But negation does just the opposite: it rechannels information into a different, but usually not totally unexpected, direction.

Leinfeller (1994, p. 79)

Negation creates an explicitly nonexistent world.

Givón (1984, p. 332)

This concluding chapter brings together some of the significant issues touched on in the discussion of negation of previous chapters and outlines the contributions made by the book to studies of negation and of *Catch-22*. To end, some suggestions are made on the implications for further research in the field.

6.1. LIMITATIONS OF TRADITIONAL THEORIES OF NEGATION

Current research on negation in functionally oriented studies seems to spread across several trends. One major line of research continues the tradition of the philosophy of language, featuring aspects such as the relation between negation and presupposition, especially metalinguistic negation (see Carston, 1996; Horn, 1989), or the supposed ambiguity of negation (see Atlas & Levinson, 1981). The research on negation in this area is extremely significant within the field of pragmatics. Researchers tend to use isolated sentences or brief exchanges as examples, however, so that the observations made on negation tend to be of a theoretical nature and do not deal with aspects that are observable only when dealing with negation in actually occurring speech. A similar limitation can be observed in speech act approaches to negation, since most research has focused on the classifications of types of denial or negative statements (see Tottie, 1991; van der Sandt, 1991; Vanderveken, 1991). Nevertheless, studies of this kind have also given rise to research on the types of negatives in actually occurring discourse, and here, Tottie's monograph on the functions of negation in English speech and writing is the best example. Tottie's (1991) work is extremely useful as a reference guide for studies on variation of negation types in speech and writing; yet, like the other approaches mentioned, it is limited when dealing with a qualitatively oriented work, in that it does not investigate in depth what it is that negation *does* in discourse. To explore what negation does in discourse, or, rather, what a speaker does when using negation in discourse, we need a cognitive approach as a point of departure.

Such a cognitive basis is found in Givón's (1978, 1979, 1984, 1993) model of negation. This model has the merit of exploring in depth the cognitive and ontological properties of negative utterances and negative lexical items, a crucial aspect when dealing with negation as a discourse phenomenon. Givón's framework is not the first and only one to deal with the cognitive and ontological properties of negation, which, of course, are topics that have been the object of study in psychology (see Clark & Clark, 1977; Wason, 1965) and philosophy (see Russell, 1905/1988), and that have been mentioned by other linguists (Leech, 1983; Lyons, 1977). Givón, however, is probably the first author to systematize the observations regarding these aspects of negation within a more elaborate linguistic framework of negation. The limitations of Givón's model have to do with the lack of specificity regarding the discourse functions of negation, in spite of the details about its cognitive properties. Thus, many authors, including Givón, point out that negation typically denies a previous assumption or expectation, or cultural knowledge. This provides a classification of the types of denial depending on the relationship between the negative and the denied affirmative proposition. This does not explain what the function of negation is *within* a stretch of discourse, however, apart from considering the relation between the negative and the affirmative terms.

I argue that Werth's (1995c/1999) text world approach to negation in discourse can provide the means of combining a cognitively and pragmatically based account of negation, along the lines described previously, with an account of the discourse functions of negation as subworld. This has to do with the organization of information in the text and with the way in which linguistic items create semantic and conceptual domains in texts. This framework is discussed in Section 6.2. together with my own development and application of the model.

When considering the approaches to syntactic and lexical negation involved in contradiction, it must be pointed out that the point of departure for most present theories is also found in philosophy (Russell, 1905/1988) and psychology (see Apter, 1982; Clark & Clark, 1977). Contradiction is not envisaged by Werth (1995c/1999), although it is mentioned as a function of negation by several authors (Escandell, 1990; Givón, 1984; Lyons, 1977; Sperber & Wilson, 1986). In this book, contradiction is approached by means of frame semantic and schema theoretic principles (see Fillmore, 1985; Pagano, 1994; and Shanon, 1981 for frame semantic approaches to negation). I proposed an application of Schank and Abelson's (1977) model and Schank's (1982) model to account for the complexities of contradiction. This approach is described in further detail in Section 6.2., together with the other functions of negation.

6.2. THE FUNCTIONS OF NEGATION IN DISCOURSE

Throughout this book, I consider the functions of negation from two different points of view: (1) the function of negation as a linguistic item within a stretch of discourse; a definition of this kind requires an analysis based on linguistic principles; and (2) the function of negation as an item within a literary work; the specification that we are dealing with fiction requires a further level of interpretation, which, I argue, can be dealt with by means of a theory of stylistics.

Starting from the functions of negation as a linguistic item, in Chapter 5 I argue for an approach to the functions of negation in discourse based on a text world model. The text world model developed by Werth (1995c/1999) takes up the asymmetricalist position of authors who, like Givón, consider negation to be marked with respect to the affirmative or positive term, and that this markedness has a cognitive basis.

Thus, negation is a natural foregrounding device typically used in discourse to deny a previous proposition that is explicitly mentioned or implicit in previous discourse. This process typically involves the defeat of an expectation. In Werth's (1995c/1999) text world theory, the definition of negation is built around the cognitive notion of subworld. As a subworld, negation contributes to the general discourse function of rechanneling or up-dating information so as to modify information previously introduced in the text world, either function-advancing propositions or world-building propositions. Within function-advancing proposi- tions, a distinction can be drawn between negation that modifies narrative propo- sitions, negation that modifies descriptive propositions, and negation that modifies information evoked from the discourse itself. In this last type, the role of frame knowledge is considered to be crucial. Thus, although frame knowledge is con- stantly activated in the processing of discourse, there are certain types of negatives that are pragmatically acceptable and recoverable only when the relevant knowledge frames affected by the negative term are adequately activated and understood. Finally, within the modification of descriptive material, a less typical function can be identified, which Werth (1995c/1999) calls *negative accommodation*. In this function, negation does not modify previously existing information but it introduces a new item to deny it. In Werth's text world model, negation is not just a static semantic notion. It emerges as a dynamic discourse process as it involves a constant reassessment of ongoing discourse.

Whereas the description of negation in these terms allows for the identification of the discourse functions of negation, that is, its contribution to the general function of organizing and updating information, the notion of subworld also encompasses the cognitive and ontological properties of negative propositions and lexical items. Thus, a subworld is defined as a conceptual and semantic domain that is triggered by a negative word. In Chapter 5 I claim that this definition could be extended to incorporate Martin's (1992) notion of *semantic prosody*, according to which the semantic influence of a linguistic item can go beyond the boundaries of the clause and stretch over a piece of discourse. In this sense, it can be said that several related negative clauses create a complex nonfactual domain that describes a complex state of affairs where something is not the case or a property fails to occur.

In addition to a function where negation updates and rechannels information by modifying information of the text world, I argue that negation also carries out a function where it blocks the flow of information, producing a sort of communicative short circuit. This is the case of a certain type of contradiction. Thus, I distinguish between three types of contradictory structures: (1) structures where the contradiction is solved by ultimately favoring one of the terms in the contradiction, as proposed by Sperber and Wilson (1986); (2) structures where the contradiction is resolvable by identifying two different conceptual domains where the contradictory terms can be said to be acceptable; and (3) structures where the contradiction is apparently irresolvable and can only be understood at a higher processing level by conveying more meaning than is manifested explicitly by means of conversational implicature. The structures of the first type can be analyzed in the same way as other negative subworlds that rechannel the flow of information. The structures of the remaining two types, however, create an apparent communicative block, which places a greater burden on the reader with regard to interpretation. In the analysis of Chapter 4, I make use of Schank's (1982) model of dynamic memory, as it provides the necessary multileveled framework with schematic structures that make it possible to understand the coexistence of apparently contradictory terms by means of identifying and understanding the different domains where each of the terms is applicable, as was observed in the contradictions of the second type.

In Chapter 5, the text world theoretical model based on Werth (1995c/1999) is complemented by Ryan's (1991b) approach to conflict in fiction. This framework provides a complementary view of the function of negation in *Catch-22* from a literary perspective. Thus, in certain parts of the discussion it is interesting to observe that negation, as a linguistic choice, is used to manifest some kind of conflict in the terms described by Ryan, either a conflict between characters' different domains, between the domain of a character and the status quo in the fictional world, or between different subdomains within one character. Although the analysis is not given in depth from this perspective, as the task is beyond the ultimate aim of the present book, it is clear that an analysis of negation is illuminating in order to understand certain types of conflict that arise in the fictional world of *Catch-22*.

The approach to negation on the lines previously described, is placed within a broader stylistic perspective to explore the connections between the functions of negation as a linguistic item and its contribution to a perceived effect. In Chapters 1 and 4, the contributions of various theories, such as systemic linguistics, pragmatics, discourse analysis, possible world theory, and schema theory, to the interpretation of literary discourse is discussed. I argue for an approach based on a cognitive reformulation of the notions of foregrounding and defamiliarization proposed by the Russian formalists, which would incorporate certain key aspects from the linguistic models discussed: the need for systematic linguistic procedures from the systemicists, the notion of a fiction as a possible world from possible world theory, and the crucial role of the reader from schema theory. The result is a model of stylistics based on rigorous linguistic analysis that will investigate the relation between the foregrounding of a linguistic feature and its perception as defamiliarizing by the reader.

In reading *Catch-22*, I focus on my perception of negation as a foregrounded feature and its contribution to the creation of a given world view that is defamiliarizing. Thus, the perception of negation as foregrounded and marked arises from the observation that the negative term was used in a systematic way in contexts where the positive or affirmative term was expected, or where the use of the negative term defeated expectations regarding the way things are standardly assumed to be. This procedure manifests a consistent defeat of expectations in the discourse of the novel, often with a humorous effect.

I point out that the functions of negation in the discourse of the novel could be classified in two main groups related to the two main functions of negatives. First, in its function of subworld that rechannels information, negation in *Catch-22* defeats expectations that arise from narrative or descriptive propositions in such a way that facts that were described as true or real are later revealed to be false or unreal. I summarize this process as one where *what is* is substituted by *what is not*, that is, a process where the appearance is given priority over the status quo of the actual fictional world. This process is interpreted as a way of deconstructing the fictional world itself, so as to lead to a questioning of the epistemic distinction between reality and illusion. In linguistic terms, this is manifested by means of a process of denial of the validity of assumptions regarding the status quo of the fictional world.

Second, in its function as a subworld that blocks the flow of information, negation is part of contradictory structures. This phenomenon, together with the recurrence of the descriptions of entities in terms of contrary lexical items, is said to be responsible for the blurring of boundaries between opposite terms, thus creating a feeling of instability in the way that the reality of the fictional world is conceptualized. The blurring of boundaries between opposites such as crazy/sane, alive/dead is ultimately interpreted as a process that seems to question the validity of linguistic classifications themselves. To sum up, the constant foregrounding of negative states and events is a defamiliarizing process where, in linguistic terms, the marked term in the polarity system, negation, paradoxically becomes the unmarked option. In more informal terms, it reflects the process described by a character in the novel as *jamais vu*, a kind of cognitive illusion where the unfamiliar acquires a strange feeling of familiarity. Finally, I argue that negation is crucial in the development of a world view that brought to the foreground certain key themes dealt with in the novel in order to question taken-for-granted assumptions about war, death, religion, economy, business, and so forth. In this sense, the world view that arises in the novel is one that is critical of a given situation, basically the reality of contemporary capitalist societies and institutionalized behavior. In Cook's (1994) terms, this world view can be understood to be schema challenging in the sense that it leads to a revision of basic aspects of our everyday experience, and can thus be schema refreshing.

6.3. FINAL REMARKS AND SUGGESTIONS FOR FURTHER RESEARCH

It can be concluded that the present book contributes to theories of negation by expanding on a previous text world model and applying this model to the analysis and interpretation of the role of negation within an extended piece of discourse. More specifically, I am concerned with those aspects of negation that have not received sufficient attention in previous models, such as the integration of the cognitive properties of negation into a theory of discourse, the need to consider negation as a dynamic process, and the incorporation of relations of contradiction into a discourse theory. Throughout the discussion in the different chapters, but especially in the analysis of negation in *Catch-22* in Chapter 5, it becomes obvious that negation within a theory of discourse needs to be accounted for with relation to the preceding discourse, in particular, with relation to the affirmative proposition that the negative operates upon and, which as Givón claims, it presupposes. This relationship between the negative proposition and the preceding propositions in discourse is shown to be particularly significant in those cases where the activation of relevant knowledge frames determined the adequate processing of the negative term, and also in those cases where the influence of one or more negative proposi-

tions was felt to stretch over a piece of discourse larger than the negative propositions themselves.

During the discussion of the different approaches to negation and in the analysis
of the data, I touch upon various interesting issues which, for reasons of space, have
not been dealt with in depth in the present work, but, which could be the point of
departure for further research. For example, a more detailed study could be carried
out regarding the types of negative lexical items and kinds of opposites, together
with their functions in discourse, an aspect that has occupied a secondary place with
regard to the analysis of the functions of syntactic negation in the present book.
Similarly, the intuitions regarding the connections between the perceived markedness of negation and a possible defamiliarizing or schema refreshing effect might
be explored further by means of an experimental study on the lines of van Peer's
(1986) investigation into the nature of foregrounding. A study of this type would
provide the means of indicating how different readers perceive and react to the
pattern of foregrounding.

There is more work to be carried out on the discourse properties of negation, and
in particular, on the view of negation as a dynamic discourse process rather than as
a static semantic feature of propositions. The present work is a proposal for a line
of research in the field of negation as a discourse unit. In this sense, the arguments
put forward in the theoretical and applied sections represent a possible point of
departure for future studies on the discoursive, pragmatic, and cognitive properties
of negation within a text world framework. As such, this model can be applied to
the analysis of negation within other discourse types, so as to develop a broader
perspective on the functions of negation, always elusive to close analysis and
systematization, but always challenging and fascinating.

Appendix

A Quantification of Negation in Catch-22

A.1. INTRODUCTION

This appendix contains three figures that provide information regarding the frequency of affixal (morphological) and nonaffixal (syntactic) negation in the novel *Catch-22* (Figure A.1); the frequency of nonaffixal negation types in *Catch-22* (Figure A.2); and the frequency of negation types in three corpora, the novel *Catch-22* and the sections of general fiction in the corpora LOB and Brown (hereafter, LOB-K, and Brown-K) (Figure A.3).

The objective of the quantification of negation is to systematize the information regarding the frequency and distribution of negation in the novel *Catch-22* and to compare the frequency of negation in this novel with the frequency of negation in two other corpora of general fiction. The comparison of the frequency of negation in *Catch-22* and the sections of fiction in the corpora LOB and Brown may provide insights regarding the role of negation as a marked feature in the novel *Catch-22*. Thus, a higher frequency of negative words in *Catch-22*, with respect to the other corpora, may be interpreted as evidence in favor of the claim that negation is a marked feature in the novel.

A.1.1. Electronic Texts Databases and Machine-Readable Corpora

Catch-22 digitized by myself; 171,110 words.

Sections on general fiction in LOB and Brown corpora (LOB-K and Brown-K); 58,000 words in each section.

A.1.2.   Corpus Analysis Tools

Micro-Concord and MonoConc.

A.1.3.   Statistics Analysis

Graphpad.

A.1.4.   Methodology

The novel was scanned to obtain a computer readable format, a search of negative words was subsequently carried out by means of the programs MicroConcord and MonoConc in the whole novel (171,110 words). Negative words are defined as what Tottie describes as "formally and semantically negative expressions" (1991, p. 7), that is, the negative words *no, not, n't, never, neither, nor, no one, none, nowhere, nobody*, and *nothing*; in addition to the words containing the negative prefixes *in-, un-, dis-, a-*, and *non-*; and the words containing the suffix *-less*, and the word *without*.

The total amount of negative words was calculated first and a distinction was established between what Tottie (1991, pp. 7–8) calls affixal (or morphological) negation and nonaffixal (or syntactic) negation. Subsequently, the frequency of syntactic negation types was calculated and compared to the frequency of syntactic negation in the sections of general fiction in the LOB and Brown corpora. The standard error was also calculated for each of the frequencies. To establish the significance of the differences between the three corpora, the chi-square test with Yates' correction was applied to each type of negation in the three corpora. The difference was considered to be significant when $p < .05$, as is standardly established.

A.2.   RESULTS AND DISCUSSION OF THE RESULTS

A.2.1.   Number and Frequency of Negative Words per 1,000 Words

The total number of negative words in the whole novel is 4,182, in a corpus of 171,110 words. The average number of negative words per 1,000 words is 24.4 which shows a frequency that is closer to the figures cited by Tottie (1982, p. 90, 1991, p. 32) for the spoken language. In Tottie's studies, the frequency of negative words per 1,000 words in the spoken language is 27.6, against 12.8 in written texts. With regard to the novel *Catch-22*, the relatively high frequency of negative words as compared to Tottie's results for written discourse can be partly explained by the fact that the novel presents a combination of narrative and dialogue. This makes it possible to have specific uses of negation that are typical of spontaneous speech, such as the pragmatic signal *no*. However, as Tottie points out (1982, p. 91), this is not the only reason for having more negative words in conversation than in writing and other factors can intervene. The discussion of these factors, however, goes beyond the purposes of the present study.

A.2.2. Frequency of Negative Words According to Grammatical Category

The frequency of affixal and nonaffixal negation types in *Catch-22* is represented in Figure A.1, and the frequency of nonaffixal (syntactic) negation types according to grammatical category is represented in Figure A.2.

Figure A.1 shows that in *Catch-22*, nonaffixal negation (79%) is more frequent than affixal negation (21%). The frequency per 1,000 words of each type is 5.2 for affixal negation and 19.3 for nonaffixal negation. In Tottie's study of two 50,000-word samples of spoken and written English (Tottie, 1991, p. 46), nonaffixal negation is significantly more frequent in the spoken sample (92%) than in the written sample (67%), while affixal negation is more frequent in the written sample (33%) than in the spoken sample (8%). These results show that the frequency of affixal and nonaffixal negation in *Catch-22* is somewhere in between what is typical for the distribution of the two negation types in Tottie's spoken and written samples.

Figure A.2 shows the frequency of syntactic negation types in *Catch-22*.

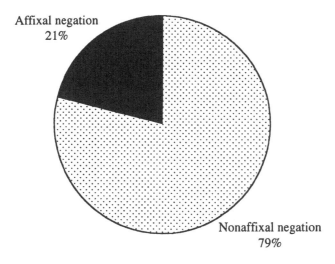

Affixal negation
21%

Nonaffixal negation
79%

FIGURE A.1.   Frequency of affixal and nonaffixal negation in *Catch-22*.

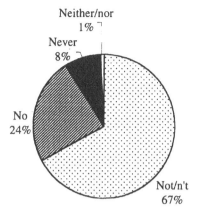

FIGURE A.2. Frequency of nonaffixal negations in *Catch-22*.

When comparing the results in Figure A.2 to those obtained by Tottie (1991, p. 194) in a study of the frequency of analytic negation (*not*) and synthetic negation (*no, never, neither, nor, none, nobody, no one, never,* and *nowhere*), it can be concluded that the data from the novel *Catch-22* show figures that are closer to what is typical in the spoken language. In Tottie's study (1991), the percentage of *not* is much higher than that of *no* in the spoken samples (67% versus 33%), while in the written samples, it is the other way around (37% versus 63%). In *Catch-22*, the percentage of *not*-negation is 67% while the percentage of *no*-negation is 33%, which shows similar results to those obtained by Tottie for the spoken samples. These results confirm Tottie's observation that "fiction, drama and poetry . . . are in different ways mimetic of the spoken mode" (Tottie, 1991, p. 9).

A.2.3. Frequency of Negation Types in Three Different Corpora

When comparing the results of the novel *Catch-22* to those obtained by searching the files containing general fiction in the corpora LOB and Brown, it is observed that the total frequency of negative items in *Catch-22* is significantly higher than in LOB and Brown. The results are shown in Figure A.3.

From the results in Figure A.3, it is observed that although the total frequency of negative words per 1,000 words in *Catch-22* is of 24.4, in LOB it is of 21.5, and in Brown it is of 16.8. The results are interpreted to be extremely significant, as $p <$ .00008 for the difference between *Catch-22* and the section of fiction in LOB, and $p <$ .00001 for the difference between *Catch-22* and Brown. The differences between the three corpora are also reflected in the general tendency of *Catch-22* to have a higher frequency of negative words in the larger categories (*not* and *affixal* negation), in spite of the deviations that are observed in categories such as *neither* and *never*. It is particularly significant that the frequency of negative words in *Catch-22* is considerably higher than the frequency of negative words in the

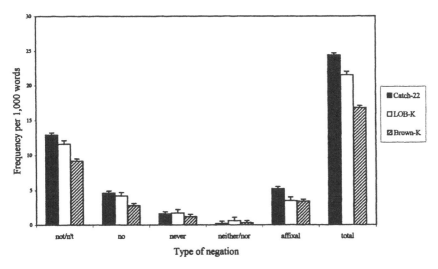

FIGURE A.3.   Frequency of negation types in three different corpora.

American corpus (Brown). Since *Catch-22* is an American novel, it can be argued that the comparison with the American corpus is more significant for the purposes of the present study than the comparison with the British corpus.

A.3.   CONCLUSIONS

Some tentative conclusions can be drawn from the quantitative analysis described in the preceding sections. With regard to the frequency of negative words in *Catch-22* as compared to the frequency of negative words in spoken and written genres analyzed by Tottie (1982, 1991), *Catch-22,* as a work of fiction, seems to share properties with spoken discourse. Indeed, the frequency of negative words in *Catch-22* is closer to the frequency of negative words in spontaneous speech than to the frequency of negative words in written discourse. Additionally, the distribution of *not-negation* and *no-negation* in *Catch-22* is also similar to the distribution of these negation types in the spoken language. The similarities of *Catch-22* with spoken discourse are partly accounted for by the fact that the novel presents a mixture of narrative and dialogue.

With regard to the comparison between the frequency of negative words in *Catch-22* and the sections of fiction in the corpora LOB and Brown, the results of the analysis show that the frequency of negation is significantly higher in *Catch-22,* especially with regard to the American corpus (Brown). As stated in the introduction to the Appendix, the higher frequency of negative words in *Catch-22* is interpreted as further evidence to the claim that negation is a marked feature in the novel.

References

Adams, M., & Collins, A. (1979). A schema-theoretical view of reading. In R. Freedle (Ed.), *New directions in discourse processing* (pp. 1–22). Norwood: Ablex Publishing Co.

Aguirre, M. (1991). Some commonsense notions on postmodernism. *Studia patricia shaw oblata, vol. III*, publicaciones de la Universidad de Oviedo. 3–13.

Allén, S. (Ed.). (1989). *Possible worlds in the humanities, arts and sciences: Proceedings of the Nobel Symposium, 65*. New York and Berlin: Mouton de Gruyter.

Allwood, J., Andersson, L. G., & Östen, D. (1974). *Logic in linguistics*. Cambridge: Cambridge University Press.

Apter, M. J. (1982). *The experience of motivation: The theory of psychological reversals*. London: Academic Press.

Atlas, J. D. (1977). Negation, ambiguity and presupposition. *Linguistics and Philosophy, 1*, 321–336.

Atlas, J. D., & Levinson, S. (1981). Negation and implicature: A problem for the standard version of radical pragmatics. In It-clefts, informativeness and logical form: Radical pragmatics. In P. Cole (Ed.), *Radical pragmatics* (pp. 32–37). New York: Academic Press.

Attardo, S. (1994). *Linguistic theories of humor*. Berlin and New York: Mouton de Gruyter.

Austin, J. K. (1962). *How to do things with words*. Oxford: Oxford University Press.

Bartlett, F. C. (1932). *Remembering*. Cambridge: Cambridge University Press.

Bateson, G. (1972). A theory of play and fantasy. In G. Bateson (Ed.), *Steps to an ecology of mind* (pp. 177–193). London: Intertext Books.

Bernárdez, E. (1995). *Teoría y epistemología del texto*. Madrid: Cátedra.

Birch, D. (1989). *Language, literature and critical practice*. London: Routledge.

Blues, T. (1971). The moral structure of *Catch-22*. *Studies in the Novel, 3*, 64–79.

Bockting, I. (1994). Mind style as an interdisciplinary approach to characterisation in Faulkner. *Language and Literature, 3*(3), 157–174.

Bolinger, D. (1977). *Meaning and form*. (pp. 37–65). London: Longman.

Bosque, I. (1980). *Sobre la negación*. Madrid: Cátedra.

Bradley, R., & Swartz, N. (1979). *Possible worlds: An introduction to logic and its philosophy*. Oxford: Blackwell.

Breuer, R. (1980). Irony, Literature and Schizophrenia. *New Literary History, 12*(1), 107–118.

Brown, G., & Yule, G. (1983). *Discourse analysis*. Cambridge: Cambridge University Press.

Brown, P., & Levinson, S. (1987). *Politeness: Some universals in language use*. Cambridge: Cambridge University Press.

Brown, R. (1973). *A first language*. Cambridge, Mass.: Harvard University Press.

Budick, S., & Iser, W. (Eds.). (1989). *Languages of the unsayable: The play of negativity in literature and literary theory*. New York/Oxford: Columbia University Press.

Burnham, C. S. (1974). Spindrift and the sea: Structural patterns and unifying elements in *Catch-22*. In J. Nagel (Ed.), *Critical essays on Catch-22*. (pp. 40–50). Encino, CA: Dickenson.

Burton, D. (1980). *Dialogue and discourse*. London: Routledge and Kegan Paul.

Burton. D. (1982). Through glass darkly: Through dark glasses. In R. Carter (Ed.), *Language and literature* (pp. 195–216). London: Allen and Unwin.

Bustos, E. (1986). *Pragmática del español: negación, cuantificación y modo*. Madrid: Universidad Nacional a Distancia.

Calvo, C. (1992). Pronouns of address and social negotiation in *As You Like it. Language and Literature, 1*(1), 5–28.

Carrell, P., Devine, J., & Eskey, D. (Eds.). (1988). *Interactive approaches to second language reading.* Cambridge: Cambridge University Press.

Carston, R. (1996). Metalinguistic negation and echoic use. *Journal of Pragmatics, 25,* 309–330.

Carter, R. A. (Ed.). (1982). *Language and literature.* London: Allen and Unwin.

Carter, R. A., & Nash, W. (1990). *Seeing through words.* Oxford: Blackwell.

Carter, R. A., & Simpson, P. (Eds.). (1989). *Language, discourse and literature.* London: Unwin Hyman.

Chapman, A. J., & Foot, H. C. (Eds.). (1976). *Humour and laughter: Theory, research and applications.* London: Wiley.

Clark, H. H. (1976). *Semantics and comprehension* (pp. 18–57). Amsterdam: Mouton de Gruyter.

Clark, H. H., & Clark, E. (1977). *Psychology and language: An introduction to psycholinguistics.* New York: Harcourt Brace Jovanovich.

Cook, G. (1994). *Discourse and literature: The interplay of form and mind.* Oxford: Oxford University Press.

Cruse, D. A. (1986). *Lexical semantics.* Cambridge: Cambridge University Press.

Davis, G. W. (1984). *Catch-22* and the Language of Discontinuity. In J. Nagel (Ed.), *Critical essays on Joseph Heller* (pp. 62–73). Boston: J. K. Hall.

de Beaugrande, R. (1980). *Text, discourse and process.* Norwood, NJ: Ablex.

de Beaugrande, R. (1987). Schemas for literary communication. In L. Halász (Ed.), *Literary discourse* (pp. 48–73). Amsterdam and Berlin: Mouton de Gruyter.

de Beaugrande, R., & Dressler, W. (1981). *Introduction to text linguistics.* London: Longman.

Derks, P., Palmer, J., Safer, E., Sherman, L., & Svebak, S. (1994). A Multidisciplinary Approach to Joseph Heller's *Catch-22*—A Symposium. International Society of Humor Studies Conference, Ithaca, New York.

Doležel, L. (1976). Narrative modalities. *Journal of Literary Semantics, 5*(1). 5–14.

Doležel, L. (1989). Possible worlds and literary fictions. In S. Allén (Ed.), *Possible worlds in the humanities, arts and sciences: Proceedings of the nobel symposium 65* (pp. 321–342). New York and Berlin: Mouton de Gruyter.

Downing, A. (1995). A functional grammar for students of English: An extended review of T. Givón (1993), *English grammar: A function-based approach. Functions of Language, 2*(2), 229–247.

Downing, A., & Locke, P. L. (1992). *A university course in english grammar.* Hemel Hempstead: Prentice-Hall.

Ducrot, O. (1972) *Dire et ne pas dire. Principes de sémantique linguistique.* Paris: Hermann.

Dummett, M. (1981). *Frege—philosophy of language.* London: The Trinity Press.

Eco, U. (1989). Report on session 3: Literature and the arts. In S. Allén (Ed.), *Possible worlds in the humanities, arts and sciences: Proceedings of the nobel symposium 65* (pp. 343–355). New York and Berlin: Mouton de Gruyter.

Emmott, C. (1994). Frames of reference: Contextual monitoring and the interpretation of narrative discourse. In M. Coulthard (Ed.), *Advances in written text analysis* (pp. 157–166). London: Routledge.

Empson, W. (1995). *Seven types of ambiguity.* Harmondsworth: Penguin. (Original work published 1930).

Enkvist, N. E. (1989). Connexity, interpretability, universes of discourse, and text worlds. In S. Allén (Ed.), *Possible worlds in the humanities, arts and sciences: Proceedings of the nobel symposium 65* (pp. 162–186). New York and Berlin: Mouton de Gruyter.

Escandell Vidal, M. V. (1990). Estrategias en la interpretación de enunciados contradictorios, *Actas del congreso de la sociedad española de lingüística, XX aniversario* (pp.923–936). Madrid: Gredos.

Fauconnier, G. (1985). *Mental spaces.* Cambridge: Cambridge University Press.

Fillmore, C. (1982). Frame semantics. In The Linguistic Society of Korea (Eds.), *Linguistics in the morning calm* (pp. 11–137). Seoul: Hanshin Publishing.

Fillmore, C. (1985). Frames and the semantics of understanding. *Quaderni di Semantica*, 6(2), 222–254.

Fish, S. (1980). *Is there a text in this class?* Cambridge, Mass.: Harvard University Press.

Fowler, R. (1977). *Linguistics and the novel.* London: Methuen.

Fowler, R. (1986). *Linguistic criticism.* Oxford: Oxford University Press.

Freud, S. (1966). *Six jokes and their relation to the unconscious.* Harmondsworth: Penguin.

Freud, S. (1976a.). La negación. In *Obras Completas, vol. 21* (pp. 253–257). Buenos Aires: Amorrotu. (1927).

Freud, S. (1976b.). El humor. In *Obras Completas, vol. 21* (pp.153–162). Buenos Aires: Amorrotu. (1927).

Garvin, P. L. (Ed.). (1964). *A prague school reader on esthetics, literary structure and style.* Washington DC: Georgetown University Press.

Gaukroger, D. (1970). Time structure in Catch-22. *Critique*, 12(2), 70–85. Reprinted in J. Nagel (Ed.), *Critical Essays on* Catch-22. Encino, CA: Dickenson.

Givón, T. (1978). Negation in language: Pragmatics, function, ontology. In P. Cole (Ed.), *Syntax and semantics 9: Pragmatics* (pp. 69–112). New York: Academic Press.

Givón, T. (1979). *On understanding grammar.* (pp. 91–142). New York: Academic Press.

Givón, T. (1984). *Syntax: A functional-typological introduction.* (pp. 321–351). Amsterdam: J. Benjamins.

Givón, T. (1989). *Mind, code and context: Essays in pragmatics.* (pp. 127–172). Hillsdale, NJ: Lawrence Erlbaum.

Givón, T. (1993). *English grammar: A function-based approach.* (pp. 187–208). Amsterdam: John Benjamins.

Goffman, E. (1974). *Frame analysis.* Harmondsworth, England: Penguin.

Greenberg, A. (1966). The Novel of Disintegration: Paradoxical Impossibility in Contemporary Fiction. *Wisconsin Studies in Contemporary Literature*, 7(1), 103–124.

Grice, H. P. (1975). Logic and conversation. In P. Cole & J. Morgan (Eds.), *Syntax and semantics 3: Speech acts* (pp. 41–57). London: Academic Press.

Hall, G. (1996). A review of G. Cook's *Discourse and Literature: The Interplay of Form and Mind.* In *Language and Literature*, 5(1), 74–77.

Halliday, M. A. K. (1973). *Exploration in the functions of language.* London: Arnold.

Halliday, M. A. K. (1978). *Language as social semiotic* (pp. 164–182). London: Arnold.

Halliday, M. A. K. (1994). *An introduction to functional grammar* (2nd ed.). London: Arnold.

Halliday, M. A. K., & Hasan, R. (1985). *Language, text and context: Aspects of language in a social semiotic perspective.* London: Arnold.

Havránek, B. (1964). The functional differentiation of the standard language. In P. L. Garvin (Ed.), *A prague school reader on esthetics, literary structure and style* (pp. 3–16). Washington DC: Georgetown University Press.

Heller, J. (1961). *Catch-22.* London: Johnathan Cape.

Hodge, R., & Kress, G. (1994). *Language as ideology* (2nd ed.). London: Routledge.

Horn, L. (1989). *A natural history of negation.* Chicago & London: The University of Chicago Press.

Huddleston, R. (1984). *An introduction to the grammar of english.* (pp. 419–432). Cambridge: Cambridge University Press.

Hunt, J. W. (1974). Comic escape and anti-vision: Joseph Heller's *Catch-22.* In J. Nagel (Ed.), *Critical essays on* Catch-22 (pp. 125–130). Encino, CA: Dickenson.

Iser, W. (1989). The play of the text. In S. Budick and W. Iser (Eds.), *Languages of the unsayable: The play of negativity in literature and literary theory* (pp. 325–339). New York and Oxford: Columbia University Press.

Iwata, S. (1998). Some extensions of the echoic analysis of metalinguistic negation. *Lingua*, 105, 49–65.

Jackendoff, R. (1983). *Semantics and cognition.* Cambridge, Mass.: MIT Press.

Jakobson, R. (1964). Closing statement: Linguistics and poetics. In T. Sebeok (Ed.), *Style in language* (pp. 350–377). Cambridge, Mass: MIT Press.

Jespersen, O. (1966). *Negation in english and in other languages* (2nd ed.). Copenhagen: Ejnar Munksgaard Publishers. (Original work published 1917).

Jespersen, O. (1961a). *A modern english grammar on historical principles. part V: Syntax, fourth volume* (pp. 426–467). London & Copenhaguen: Allen & Unwin and Ejnar Munksgaard. [1924]

Jespersen, O. (1961b). *A modern english grammar on historical principles. part VI: Morphology* (pp. 464–489). London & Copenhaguen: Allen & Unwin and Ejnar Munksgaard. [1924].

Jordan, M. P. (1998). The power of negation in English: Text, context and relevance. *Journal of pragmatics, 29,* 705–752.

Just, M. A., & Clark, H. H. (1973). Drawing inferences from the presuppositions and implications of affirmative and negative sentences. *Journal of Verbal Learning and Verbal Behavior, 12,* 21–31.

Keegan, B. M. (1978). *Joseph Heller: A reference guide.* Boston: G.K. Hall.

Kempson, R. (1975). *Presupposition and the delimitation of Semantics.* Cambridge: Cambridge University Press.

Klima, E. S. (1964). Negation in English. In J. A Fodor & J. J. Katz (Eds.), *The structure of language: Readings in the philosophy of language.* (pp. 246–323). Hemel Hempstead: Prentice Hall.

Koestler, A. (1966). *The act of creation.* London: Pan Books.

Krassner, P. (1993). An impolite interview with Joseph Heller. In A. J. Sorkin (Ed.), *Conversations with Joseph Heller* (pp. 6–29). Jackson: University Press of Mississippi.

Kripke, S. (1971). Semantic considerations of modal logic. In L. Linsky (Ed.), *Reference and modality* (pp. 63–73). Oxford: Oxford University Press.

Kuno, S. (1993a, December). Negation and Extraction. Paper presented at the 'Seminario de Semántica Cognitiva y Gramática Funcional, University of Seville, 1993.

Kuno, S. (1993b, December). Remarks on Negative Islands. Paper presented at the 'Seminario de Semántica Cognitiva y Gramática Funcional, University of Seville. 1993.

Kurrick, M. J. (1979). *Literature and negation.* New York: Columbia University Press.

Labov, W. & Fanshel, D. (1977). *Therapeutic discourse.* (pp. 334–343). New York: Academic Press.

Lakoff, G. (1989). Some empirical results about the nature of concepts. In *Mind and Language, 4*(1 & 2), 103–129.

Langacker, R. W. (1987). *Foundations of cognitive grammar: Vol. I, theoretical prerequisites.* Stanford, CA: Stanford University Press.

Langacker, R. W. (1991). *Foundations of cognitive grammar: Vol. II, descriptive application.* (pp. 132–141). Stanford: Stanford University Press.

Leech, G. (1974). *Semantics.* Harmondsworth: Penguin.

Leech, G. (1983). *Principles of pragmatics.* London: Longman.

Leech, G., & Short, M. (1981). *Style in fiction.* London: Longman.

Leinfeller, E. (1994). The broader perspective of negation. *Journal of Literary Semantics, 13*(2), 77–98.

Levinson, S. (1983). *Pragmatics.* Cambridge: Cambridge University Press.

Lewis, D. (1979). Possible worlds. In M. J. Loux (Ed.), *The possible and the actual: Readings in the metaphisics of modality* (pp. 182–189). Ithaca and London: Cornell University Press.

Louw, B. (1993). Irony in the text or insincerity in the writer? The diagnostic potential of semantic prosodies. In M. Baker, G. Francis, & E. Tognini-Bonelli (Eds.), *Text and technology* (pp. 157–176). Philadelphia and Amsterdam: John Benjamins.

Loux, M. J. (1979). *The possible and the actual: Readings in the metaphisics of modality.* Ithaca and London: Cornell University Press.

Lyons, J. (1977). *Semantics.* Cambridge: Cambridge University Press.

Lyons, J. (1981). *Language, meaning and context.* Suffolk: Fontana.

Lyons, J. (1995). *Linguistic semantics.* Cambridge: Cambridge University Press.

Marsh, R. C. (Ed.). (1988). *Bertrand Russell: Logic and knowledge.* London: Unwin Hyman.

Martin, J. R. (1992). *English text: System and structure.* Philadelphia/Amsterdam: John Benjamins.

McCawley, J. D. (1981). *Everything that linguists wanted to know about logic but were afraid to ask.* Oxford: Blackwell.

McCawley, J. D. (1995). Jespersen's 1917 monograph on negation. *Word, 46*(1), 29–40.

McHale, B. (1987). *Postmodernist fiction*. London: Methuen.

Mellard, J. M. (1968). *Catch-22*: Déjà vu and the Labyrinth of Memory. *Bucknell Review, 16*(2), 29–44.

Merrill, R. (1987). *Joseph Heller*. Boston, Mass.: Twayne Publishers.

Merrill, S. (1993). *Playboy* Interview. In A. J. Sorkin (Ed.), *Conversations with Joseph Heller*. (pp. 144–176). Jackson: University Press of Mississippi.

Miall, D. S., & Kuiken, D. (1994). Beyond text theory: Understanding literary response, *Discourse Processes, 17*, 337–352.

Miall, D. S., & Kuiken, D. (1998). The form of reading: Empirical studies of literariness. *Poetics, 25*, 327–341.

Minsky, M. L. (1975). A framework for representing knowledge. In P. Winston (Ed.), *The psychology of computer vision* (pp. 211–227). New York: McGraw-Hill.

Mukarowsky, J. (1964). Standard language and poetic language. In Paul L. Garvin (Ed.), *A prague school reader on esthetics, literary structure and style* (pp. 17–30). Washington, DC: Georgetown University Press.

Müske, E. (1990). Frame and literary discourse, *Poetics, 19*, 433–461.

Nagel, J. (1974a), *Catch-22* and Angry Humor: A Study of the Normative Values of Satire. *Studies in American Humor, 1*, (2), 99–106.

Nagel, J. (1974b) *Critical essays on* Catch-22. Encino, CA: Dickenson.

Nagel, J. (1984). *Critical essays on Joseph Heller*. Boston: J. K. Hall.

Nash, W. (1985). *The language of humour*. London: Longman.

Nelson, T. A. (1971). Theme and structure in *Catch-22*. *Renascence, 23*(4), 173–182.

Norrick, N. (1986). A frame-theoretical analysis of verbal humor: Bisociation as schema conflict. *Semiotica* 60(3-4), 225–245.

Norrick, N. (1993). *Conversational joking*. Bloomington: Indiana University Press.

Oh, C., & Dinneen, D. A. (Eds.). (1979). *Syntax and semantics 11: Presupposition*. New York: Academic Press.

Olson, D. R. (1997) The written representation of negation, *Pragmatics and Cognition, 5*(2), 235–252.

Pagano, A. (1994). Negatives in written text. In M. Coulthard (Ed.), *Advances in Written Text Analysis* (pp. 250–265). London: Routledge.

Partee, B. H. (1989). Possible worlds in model-theoretic semantics: A linguistic perspective. In S. Allén (Ed.), *Possible worlds in the humanities, arts and sciences: Proceedings of the nobel symposium 65* (pp. 93–123). New York and Berlin: Mouton de Gruyter.

Pavel, T. (1985). Literary narratives. In T. A. van Dijk (Ed.), *Discourse and literature* (pp. 85–103). Amsterdam/Philadelphia: John Benjamins.

Pavel, T. (1986). *Fictional worlds*. Cambridge and London: Harvard University Press.

Petöfi, J. (1989). Possible worlds—Text worlds: Quo vadis linguistica? In S. Allén (Ed.), *Possible worlds in the humanities, arts and sciences: Proceedings of the nobel symposium 65* (pp. 209–218). New York and Berlin: Mouton de Gruyter.

Petrey, S. (1990). *Speech acts and literary theory*. London: Routledge.

Pilkington, A. (1996). Relevance theory and literary style, *Language and Literature, 5*(3), 157–163.

Pinsker, S. (1991). *Understanding Joseph Heller*. Columbia: University of South Carolina Press.

Pratt, M. L. (1977). *Towards a speech act theory of literary discourse*. Bloomington: Indiana University Press.

Protherough, R. (1971). The sanity of *Catch-22*. *The Human World, 3*, 59–70.

Quirk, R., Greenbaum, S., & Svartvik, J. (1985). *A comprehensive grammar of the english language* (pp. 775–799). London: Longman.

Quirk, R., & Greenbaum, S. (1990). *A student's grammar of the english language* (pp. 223–230). London: Longman.

Ramsey, V. (1968). From here to absurdity: Heller's *Catch-22*. In T. B. Whitbread (Ed.), *Seven contemporary authors: Essays on Cozzens, Miller, West, Golding, Heller, Albee, and Powers* (pp. 99–118). Austin and London: University of Texas Press.

Raskin, V. (1985). *Semantic mechanisms of humour.* Dordrecht/Boston: Reidel Publishing Co.

Rescher, N. (1979). The ontology of the possible. In M. L. Loux (Ed.), *The possible and the actual: Readings in the metaphisics of modality* (pp. 166–181). Ithaca and London: Cornell University Press.

Rosenhan, D. L. (1973). Estar sano en lugares insanos. *Science, 179,* 250–258.

Rosch, E. (1973). Natural Categories. *Cognitive Psychology, 4,* 328–506.

Ruderman, J. (1991). *Joseph Heller: Criticism and interpretation.* New York: Continuum Publishing Company.

Rumelhart, D. E. (1980). Schemata: The building blocks of cognition. In R. J. Spiro, B. Bruce, & W. Brewer (Eds.), *Theoretical issues in reading comprehension* (pp. 33–58). Hillsdale NJ: Lawrence Erlbaum.

Rumelhart, D. E., and Norman, D. A. (1981). Analogical processes in learning. In J. R. Anderson (Ed.), *Cognitive skills and their acquisition* (pp. 335–360). Hillsdale, NJ: Lawrence Erlbaum.

Russell, B. (1988). On denoting. In R. C. Marsh (Ed.), *Bertrand Russell: Logic and knowledge* (pp. 39–57). London: Unwin Hyman. (Original work published 1905).

Ryan, M. L. (1985). The modal structure of narrative universes, *Poetics Today, 6*(4), 717–755.

Ryan, M. L. (1991a). Possible worlds and accessibility relations: A semantic typology of fiction. *Poetics Today, 12*(3), 553–576.

Ryan, M. L. (1991b). *Possible worlds, artificial intelligence and narrative theory.* Bloomington: Indiana University Press.

Sacks, H. (1974). An analysis of the course of a joke's telling in conversation. In R. Bauman and J. Sherzer (Eds.), *Explorations in the ethnography of speaking* (pp. 337–353). Cambridge: Cambridge University Press.

Schank, R. C. (1982). *Dynamic memory.* Cambridge: Cambridge University Press.

Schank, R. C., & Abelson, R. (1977). *Scripts, plans, goals and understanding.* Hillsdale, New Jersey: Lawrence Erlbaum.

Searle, J. R. (1969). *Speech acts.* Cambridge: Cambridge University Press.

Searle, J. R. (1975). The logical status of fictional discourse. *New Literary History, 6,* 319–332.

Seed, D. (1989). *The Fiction of Joseph Heller: Against the Grain.* New York: St.Martin's Press.

Semino, E. (1993). A Review of Ryan's (1991). *Possible Worlds, Artificial Intelligence and Literary Theory. Language and Literature, 2*(2), 146–148.

Semino, E. (1994). *Poems, Schemata and Possible Worlds: Text Worlds in the Analysis of Poetry.* Unpublished doctoral dissertation, Lancaster, Lancaster University.

Semino, E. (1995). Schema Theory and the Analysis of Text Worlds in Poetry, *Language and Literature,4,* 2, 79–108.

Semino, E. (1997) *Language and world creation in poems and other texts.* London: Longman.

Shanon, B. (1981). What is the frame?—Linguistic indications. *Journal of Pragmatics, 5,* 35–44.

Shklovsky, V. B. (1965). Art as technique. In L. T. Lemon and M. J. Reis (Eds.), *Russian formalist criticism: Four essays* (pp. 3–24). Lincoln: University of Nebraska Press. (Original work published 1917)

Short, M. H. (1989). *Reading, analysing and teaching literature.* London: Longman.

Short, M. H. (1995). Understanding conversational undercurrents in 'The Ebony Tower' by John Fowles. In P. Verdonk and J. J. Weber (Eds.), *Twentieth-century fiction: From text to context.* (pp. 45–62). London: Routledge.

Short, M. H. (1996). *Exploring the language of poems, plays, and prose.* London: Longman.

Shulz, T. R. (1976). A cognitive-developmental analysis of humour. In A. J. Chapman and H. C. Foot (Eds.), *Humour and Laughter: Theory, Research and Applications* (pp. 11–36). London: Wiley.

Simpson, P. (1989). Politeness phenomena in Ionesco's *The Lesson*. In R. Carter and P. Simpson (Eds.), *Language, discourse and literature* (pps. 171–194). London: Unwin Hyman.

Simpson, P. (1993). *Language, ideology and point of view*. London: Routledge and Kegan Paul.

Solomon, E. (1969). From Christ in Flanders to *Catch-22*: An approach to war fiction. *Texas Studies in Literature and Language, 11*(1), 852–866.

Solomon, J. (1974). The Structure of Joseph Heller's *Catch-22*. In J. Nagel (Ed.), *Critical Essays on Catch-22*. (pp. 78–88). Encino, CA: Dickenson.

Sperber, D., & Wilson, D. (1986). *Relevance*. Oxford: Blackwell.

Tannen, D. (Ed.). (1993). *Framing in discourse*. Oxford: Oxford University Press.

Tanner, T. (1971). *City of words: American fiction 1950–1970*. New York/London: Harper and Row Publishers.

Teleman, U. (1989). The world of words—and pictures. In Allén (Ed.), *Possible worlds in the humanities, arts and sciences: Proceedings of the nobel symposium 65* (pp. 199–208). New York and Berlin: Mouton de Gruyter.

Thorndyke, P. W., & Yekovich, F. R. (1980). A critique of schema-based theories of human story memory. *Poetics, 9*, 23–49.

Toolan, M. (Ed.). (1992). *Language, text and context: Essays in stylistics*. London: Routledge.

Tottie, G. (1982). Where do negative sentences come from? *Studia Linguistica, 36*(1), 88–105.

Tottie, G. (1991). *Negation in english speech and writing: A study in variation*. San Diego: Academic Press.

Tucker, L. (1984). Entropy and information theory in *Something Happened*. *Contemporary Literature, XXV*(3), 323–340.

van der Sandt, R. A. (1991). Denial. *Papers from the 27th Regional Meeting of the CLS: Parasession on Negation, CLS, 27*, 331–344.

Vanderveken, D. (1991). *Meaning and speech acts*. Cambridge: Cambridge University Press.

van Dijk, T. A. (1977). *Text and context: Explorations in the semantics and pragmatics of discourse*. London: Longman.

van Dijk, T. A. (Ed.). (1982). *New Developments in Cognitive Models of Discourse Processing* (special issue of *Text 2-1/3*). Amsterdam: Mouton de Gruyter.

van Dijk, T. A. (1985). *Discourse and literature*. Amsterdam and Philadelphia: John Benjamins.

van Dijk, T. A., & Kintsch, W. (1983). *Strategies of discourse comprehension*. New York: Academic Press.

van Peer, W. (1986). *Stylistics and psychology: Investigations of foregrounding*. London: Croom Helm.

van Peer, W. (1988). *The taming of the text*. London: Routledge.

Verdonk, P., & Weber, J. J. (Eds.). (1995). *Twentieth-century fiction: From text to context*. London: Routledge.

Volterra, V., & Antinucci, F. (1979). Negation in child language: A pragmatic study. In E. Ochs & B. B. Schieffelin (Eds.), *Developmental pragmatics* (pp. 281–292). London: Academic Press.

Waldmeir, J. J. (1964). Two novelists of the absurd: Heller and Casey. In J. Nagel (Ed.), *Critical essays on Catch-22*. (pp. 150–154). Encino, CA: Dickenson.

Walsh, J. (1982). *American war literature, 1914 to Vietnam*. New York: St. Martin's Press.

Wason, P. C. (1965). The contexts of plausible denial. *Journal of Verbal Learning and Verbal Behavior, 4*, 7–11.

Weixlmann, J. (1974). A Bibliography of Joseph Heller's *Catch-22*. *Bulletin of Bibliography, 31*(1), 32–35.

Werth, P. (1976). Roman Jakobson's verbal analysis of poetry. *Journal of Linguistics, 12*, 21–73.

Werth, P. (1984). *Focus, coherence and emphasis*. London: Croom Helm.

Werth, P. (1993). Accommodation and the myth of presupposition: The view from discourse, *Lingua, 89*, 39–95.

Werth, P. (1994). Extended metaphor—A text-world account. *Language and Literature, 3*(2), 79–104.

Werth, P. (1995a). How to build a world (in a lot less than six days, using only what's in your head). In K. Green (Ed.), *New essays in deixis* (pp. 49–80). Amsterdam: Rodopi.

Werth, P. (1995b). 'World enough, and time': Deictic space and the interpretation of prose. In P. Verdonk and J. J. Weber (Eds.), *Twentieth-century fiction: From text to context.* (pp. 181–205). London: Routledge.

Werth, P. (1995c) *Text worlds: Representing conceptual space in discourse.* Pre-publication manuscript copy. (Published in 1999 in London by Longman)

Widdowson, H. (1975). *Stylistics and the teaching of literature.* London: Longman.

Widdowson, H. (1992). *Practical stylistics.* London: Longman.

Winston, P. (1975). *The psychology of computer vision.* New York: McGraw-Hill.

Author Index

Subject Index

Lightning Source UK Ltd.
Milton Keynes UK
UKOW06f2005100715

254982UK00003B/69/P